Elizabeth Doerr

BRIDGING ART AND MECHANICS

THE UNABRIDGED STORY OF CORUM'S GOLDEN BRIDGE

watchprint.com
publications on jewellery and watches

This book is dedicated to the memory of René Bannwart,
who has never failed to
utterly delight those who look for the fantastical
in the mundane and then strap it to their wrists.

FOREWORD

In China, there is a renowned mountain by the name of Mount Kunlun, which is as extraordinary as Europe's Mont Blanc. As it so happens, in the Chinese translation of Corum, the character and tonal pronunciation are exactly the same as Kunlun. It is my dream to make Corum timepieces as extraordinary as Mount Kunlun.

When I first heard about the Golden Bridge, I could not grasp the purity of this timepiece. Upon seeing it, I fell in love with it immediately.

Not only is the Golden Bridge a Corum icon, it is also a *haute horlogerie* masterpiece of the highest purity. The Golden Bridge, as a unique creation in the world of watchmaking, perpetuates the historical legitimacy of the brand with its baguette-shaped movement and hand-engraved plate and bridges made of 18-karat gold. As a timepiece for connoisseurs, collectors, and lovers of high watchmaking, it plays an integral part in Corum's unique history. Even after thirty-five years, it remains an icon—not only for the brand but for the entire industry.

This unique timepiece is beyond a doubt a symbol of the distinction, audacity, and passion of Corum, and it is with this in mind that I would like to share with you this book dedicated to the Golden Bridge, one of the most timeless and exclusive watches ever made.

Hon Kwok Lung,
Chairman, Citychamp Watch & Jewellery Group Limited

ISBN 978-2-940506-08-8

Publishing Managers: Suzanne Wettstein and Fabrice Mugnier

Author: Elizabeth Doerr

Copy editor: Ashley Benning

Photo shoot manager: Sherry Williams Pantieri

Photography: Jack Guy, Ian Skellern, Robert-Jan Broer, Elizabeth Doerr and the archives of Corum and the Musée International d'Horlogerie (MIH)

Design: Jan Abellan, Andeers Sàrl

Printing: IRL plus SA, Renens, Switzerland
Printed and bound in Switzerland

printed in
switzerland

See our publications on watches and jewelry on our website www.watchprint.com.

TABLE OF CONTENTS

I

THE BIRTH OF A CREATIVE HOROLOGICAL ALTERNATIVE

THE BIRTH OF A CREATIVE HORLOGICAL ALTERNATIVE

In 1350, René Bannwart's ancestors left what would later become the German state of Bavaria and settled not terribly far away in Switzerland. The family's coat of arms, which contains three fir trees, reflects the meaning of their last name: "forest and field warden" in southern German dialect and "forester" in Old High German.

"I was particularly lucky because the walls of Basel concealed for me treasures of the history of art and marvelous examples of beautiful proportions."

Bannwart himself entered the world more than five centuries later on March 16, 1915 in Zurich. His family was no longer involved in forestry; his father was in the business of printing. Because of this, they moved around the country a bit, and René spent his formative years in the city of Basel. These early surroundings would serve him well. Basel, after all, hosts the most important watch fair in the world, known today as Baselworld, which can literally determine the outcome of a watch manufacturer's future.

Few visitors to the fair, however, take the time to learn more about Basel, a cultivated, historic city imbued with art, architecture, and culture. And it is here in these fairy tale-like surroundings that Bannwart spent his youth, years that would shape him into the cultured, artistic, creative soul he later proved to be. In fact, he said in an interview with *Europa Star* at the Basel Fair of 1963, "Though my artistic points of view are naturally innate (either you

Left: A formative location for René Bannwart: Basel's Tinguely fountain

Right: Another formative location for René Bannwart: Geneva

Opposite: Corum Caliber 13 measured just 30.5 mm at its longest point.

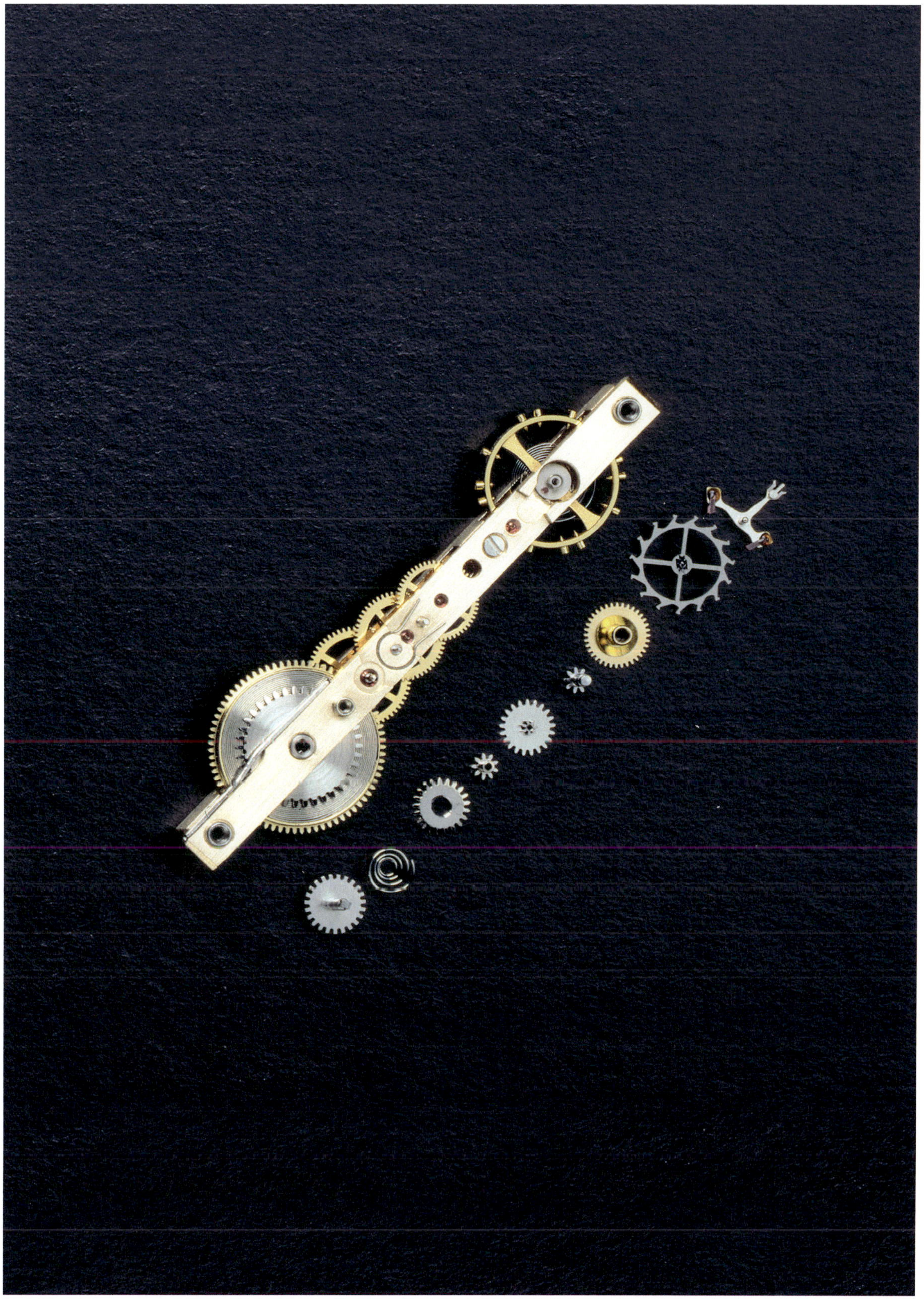

Variations on early Golden Bridge models, including a special Spatiale containing a movement shaped like the letter H have them or you don't), it's undeniable that the environment here continues to contribute to developing and orienting myself in a well-determined direction. I was particularly lucky because the walls of Basel concealed for me treasures of the history of art and marvelous examples of beautiful proportions. Later, completing this first molding of taste, was Geneva, the other site of majestic art, where I finished my commercial studies."

The fact that Bannwart's family moved around quite a bit—which was unusual in Europe at the time—probably turned René into a well-adjusted, flexible, and open-minded type of person.

From left to right
Jean-René Bannwart,
Simone Ries,
and René Bannwart

He was quite artistically inclined, saying later that if he had not gotten involved in horology, he would have wanted to become an architect. He was particularly enamored of the aesthetic architecture of Basel's old town, near the cathedral, with the Rhine passing right through its center. The local art museum also particularly enthralled him, and he cited the impressionist period as perhaps his favorite and his favorite painters being Jean-Baptiste-Camille Corot, Pierre-Auguste Renoir, and Jean-Antoine Watteau. He also came to enjoy more modern painters, in particular early Picasso, Marc Chagall, Gustav Klimt, and Georges Braque in addition to the sculptures of Henry Moore, Giò Pomodoro, and Jean Tinguely, the last a Swiss painter and sculptor (1925–1991), who also grew up in Basel.

The fact that Bannwart's family moved around quite a bit—which was unusual in Europe at the time—probably turned René into a well-adjusted, flexible, and open-minded type of person. The family landed in Geneva in time for him to complete his studies in business and economics. Though he and his parents spoke French at home, he continued with German at school, thereby mastering both the main horological languages as mother tongues.

He finished his studies at the École Supérieure de Commerce, the commercial school in Geneva, in the period following the global economic crash of 1929, earning a *maturité commercial* just as the worst economic crisis the world had ever seen was in full swing. As there was no work to be had, he remained at school one more year just to learn a few more languages, including English, Italian, and Spanish, providing him with mastery of the most important communication tools needed for doing business.

Bannwart may not have known it at the time, he had stumbled upon the path leading to his calling.

Finding a job continued to be difficult, so eventually he took one as a bicycle messenger for Patek Philippe. *Glück im Unglück*, as a German-speaker might say: the poor circumstances provided for this lucky coincidence, and though Bannwart may not have known it at the time, he had stumbled upon the path leading to his calling.

The Golden Bridge was sometimes a transparent dream in 18-karat yellow gold and diamonds.

FROM PATEK PHILIPPE TO OMEGA

In 1933, at the tender age of eighteen, Bannwart finally came into contact with the world of luxury watchmaking. Patek Philippe proved to be an excellent training ground for the studied businessman, even if his first task was less than glamorous: he delivered things, collected things, and bided his time until needed in the room where the secretaries worked. One day, a secretary lamented having to write a business letter in German, at which point Bannwart—who obviously knew his way around formal missives in the Teutonic language—offered to help. This small act edged him further down the right path.

Patek Philippe proved to be an excellent training ground for the studied businessman, even if his first task was less than glamorous.

Because of his now-disclosed familiarity with foreign languages, he was asked to join the Patek Philippe sales team in the shop "downstairs"—a word that alluded to the famous "salon" on Rue du Rhône that overlooks Lake Geneva. This led to a post as a traveling salesman in Scandinavia and southern Europe, through which he developed good relationships with Patek Philippe's clients. He also honed his design skills working on window displays and organizing fashion shows and exhibitions.

His job seemed stable, and Bannwart married in 1938. But then World War II began in 1939, a circumstance that would lead to the end of Bannwart's seven-year career at Patek Philippe, as the luxurious watchmaker—which was creating masterpieces like the Packard and the Supercomplication in this era—needed to cut back on staff to weather the conflict. Unlike some of its contemporaries, Patek Philippe did not manufacture timepieces for the armed forces. Fifty percent of the staff was let go according to a document Bannwart wrote in 1997 tracing the origins of Corum. This document was actually a letter addressed to Corum's agents in Hong Kong at the end of the 1990s, Jack Cua and Carrie Tsang. They had asked to better understand the story of Corum's beginnings, and René obliged, writing several pages in French.

Bannwart and his wife moved to the bilingual city of Biel—which French-speakers call Bienne—where he began working for Omega in 1940 with the war still underway. Biel is not only the seat of today's Swatch Group, which now owns Omega and seventeen other brands, but also a city rich in watchmaking culture and suppliers.

Thanks to the shrewd leadership of the by all accounts exceptional Adolphe Vallat, Bannwart was able to set up something at Omega that had not yet seen the light of day there—or in any other Swiss manufacture at that time for that matter: something that might be called the "creation department." Though it might seem hard to believe, this was an absolute first in Swiss watchmaking; up to that point watch companies did not truly "create" their products the same way they do today, but rather left the look of their timepieces to the designs that the case and dial makers brought with them. In that period, case and dial makers were always outside suppliers, which is why watches of one era look terribly similar to each other. *Omega Saga*, an encyclopedic work by former Omega museum director Marco Richon, goes so far as to say that the success of implementing Vallat's ideas were what led Omega down its prosperous commercial path.

Designers in the modern sense of the word did not exist as such in the watch industry of the 1940s, leaving Vallat and Bannwart's new vision to establish a novel way of creating wristwatches. Part of Vallat's idea was to maintain good contact with the brand's global customers, which meant that Bannwart had to remain in dialogue with foreign clients and create specific ideas for various markets. Bannwart also continued to design the window displays for Omega at the Basel Fair, just as he had done for Patek Philippe when he worked there.

Designers in the modern sense of the word did not exist as such in the watch industry of the 1940s, leaving Vallat and Bannwart's new vision to establish a novel way of creating wristwatches.

And, perhaps not so surprisingly when one considers his apparent talent, Richon writes that in 1948 Bannwart was the first to adopt the idea of applying luminous substance by putting a dot of radium into a recess to make a luminous dot marker. While this doesn't seem special today, as a great many elements like numerals, markers, and logos are now luminous on the dials of luxury watches, in the 1940s this had not yet been instituted. The technique quickly spread to become a luxury standard. That same year, Bannwart became the official head of the "creation" department at Omega, thereby taking *haute horlogerie* a step further into the future by introducing something new: a sense of design and a sense of creation from inside the factory.

FATHER OF THE OMEGA FLAGSHIPS CONSTELLATION AND SEAMASTER

Remaining at Omega for fifteen years, Bannwart is considered the "father" of the epochal Seamaster line introduced in 1948. The direct aftermath of the war inspired a time of highly utilitarian design thanks to experience garnered from the watches Omega delivered to the British armed forces, with the result of the accumulated experience united in the water-resistant, reliable, robust, automatic Seamaster. In fact, the Seamaster could almost have been called a civilian version of some of the war's most appreciated timepieces. It was armored against moisture by an O-ring, a hermetically sealed crown, and a hard-to-break hesalite crystal. One funny thing about the Seamaster is that according to Richon, Vallat was in a huge hurry to get it on the market, but Bannwart was never really satisfied with what he was coming up with, finding the designs still too heavy. When Vallat prodded more and Bannwart with misgivings showed him what he had accomplished, Richon writes that Vallat exclaimed, "But that's exactly what I needed!"

Bannwart is considered the "father" of the epochal Seamaster line introduced in 1948.

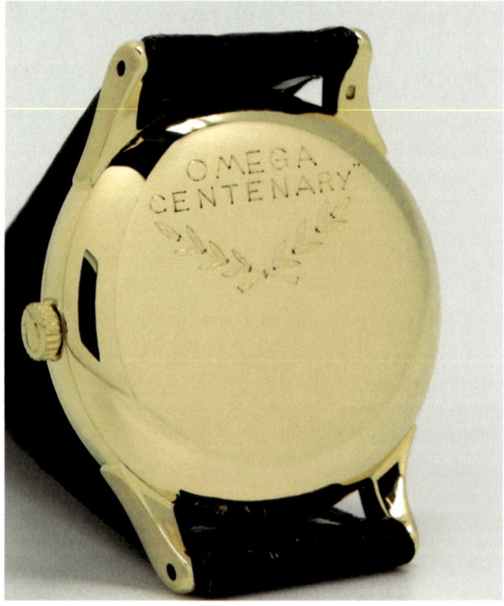

Left: An original Omega Centenary from 1948, but recently photographed (photo courtesy Robert-Jan Broer)

Right: The back of the original Omega Centenary from 1948 complete with engraving (photo courtesy Robert-Jan Broer)

As Europe recovered from the conflict, Bannwart found the way to his preferred elegance and authored some of Omega's most refined models, beginning with the Centenary line. The Centenary of 1948, an automatic wristwatch chronometer, was created to celebrate Omega's 100th anniversary (hence the name). Its premise was to unite the sum of experience acquired in 100 years of producing watches. It was released in two sizes limited to 2,000 and 4,000 pieces, respectively, and though it remained utilitarian and minimalist in its dial clarity, one version of its elegant gold case including whimsical lugs resembling the decorative neck of a violin was the thinnest of its era.

In 1954, Bannwart was named head of production and became responsible for Omega's entire collection.

Reacting to the demand it created, Omega released its successor, the Constellation, four years later in 1952. The Constellation was the absolute Omega flagship well into the 1970s and today remains one of the brand's more playful evergreens. It also marked the brand's first serially manufactured wrist chronometers. "Created by René Bannwart and produced by Merusa in Biel, it became a sensation by deliberately breaking with the uniformity and the monotony of the dials of the era," Richon wrote. In 1954, Bannwart was named head of production and became responsible for Omega's entire collection.

An original Omega Constellation from 1952, recently photographed (photo courtesy Robert-Jan Broer)

Bannwart's ideals were based on the concept that a watch was made to be seen above all, and that the mechanics, which nonetheless had to work reliably, were just one part of a whole. Richon, in fact, wrote that 1953 was the year that Omega released its most elegant creations ever, with the presentation and finishing particularly luxurious. Indeed, Bannwart also worked on automatic ladies' watches at Omega, including the interesting Sapphette, whose sapphire crystal facets were designed to evoke the brilliance of a diamond.

THE CRAZY COOL CORUM COLLECTION

Before the Golden Bridge, Corum had already established several collection pillars; the Admiral's Cup and the Romulus would go on to become evergreens that remain in the brand's collection to this very day. The Buckingham was a square watch with typical 1950s design, even though it launched in 1965. Its big size—hardly noticeable due to the clever curvature of the case made of just 35 grams of solid gold—gave it a fresh look letting consumers know that it was not from the fifties, but rather the here and now.

Prior to the Golden Bridge, Corum also sort of specialized in designs that constituted out-of-the-ordinary wristwatches and were internally categorized as "specials." One of these was naturally the Coin Watch, which saw the light of day in 1964. Advertised using the tagline "time in money," the concept of the Coin Watch was so radically different from everything else produced in the swinging sixties that it couldn't help but attract a great deal of attention. The top half of the coin posed as the dial while the company's craftsmen placed an ultrathin movement inside the authentic Double Eagle $20 gold coin, which had been carefully sawn in half.

Inspired by an automobile whose image projects tradition, luxury, and exclusivity, the Corum Rolls-Royce is a timepiece whose radiator grille design clearly echoes that of its namesake: the Rolls-Royce automobile. The first of these models available in both white and yellow gold was shown at traditional London jeweler Garrard in 1976. The grille hid a sapphire crystal, behind which a beautifully engraved and skeletonized, ultra-flat movement could be seen: Frédéric Piguet Caliber 9.21.

The case of this 1965 Buckingham model is crafted from only 35 grams of finely brushed gold; the dial comprises spelled-out numbers.

Above: Inspired by the Rolls-Royce Silver Shadow, the Corum Rolls-Royce is characterized by a striking grille hiding an ultrathin skeletonized movement.

Right: The authentic American $20 Double Eagle gold coin forms the basis, and poses as the dial, of Corum's signature Coin Watch.

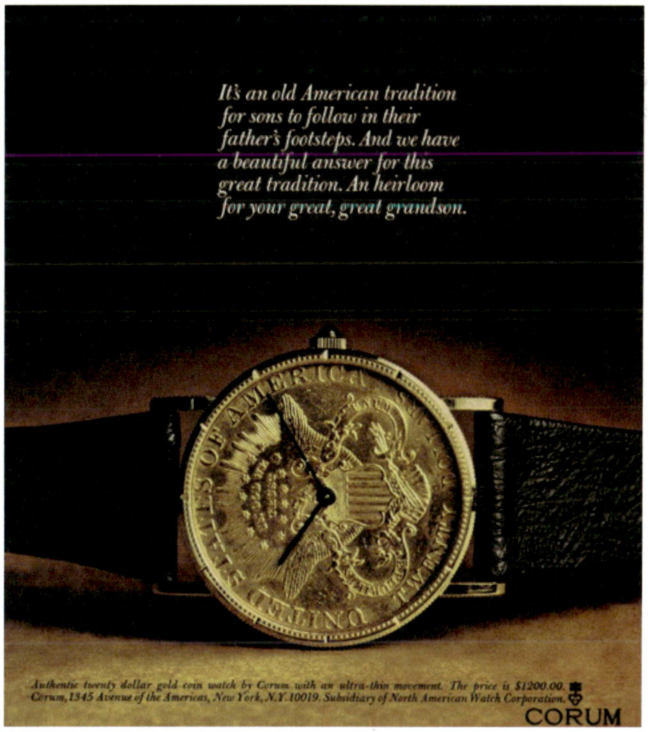

A QUORUM AS A BASE FOR A NEW COMPANY

In 1954, Bannwart was thirty-nine years old. He had had experienced much success, but it was time for him to make his own mark on Swiss watchmaking. "I had always dreamed of becoming independent one day," he wrote in the 1997 letter to Cua and Tsang.

Bannwart had experienced much success, but it was time for him to make his own mark on Swiss watchmaking.

His uncle, Gaston Ries, was a watchmaker and already owned a small factory in La Chaux-de-Fonds that he had founded in 1924. Bannwart's own father came from La Chaux-de-Fonds as did his wife, so the family had a strong connection to Switzerland's third-largest French-speaking city.

Ries did not produce watches under his own name, but rather the brand name Endura, and acted as what we would most likely understand today as a private labeler. A private labeler manufactures timepieces for someone else, practically functioning as the middleman between the company or brand selling the watches and the manufacturing process. (In the 1970s, as the quartz crisis was dawning, Endura—which was separate from the co-owned Ries, Bannwart & Co.—was assimilated by ASUAG, which, as watch history teaches us, was destined by mergers, acquisitions, and more assimilation to become a major part of the Swatch Group; Ries had passed away in 1958.)

Endura also worked as an assembler and finisher of movements (known in French as *terminage*), which it did for many other brands, including some globally known ones. Demand for watches was great in the aftermath of World War II, and Ries did good business. But by the mid-1950s, the demand had begun to decrease, and Ries was trying to think of ways to invigorate his business again. One way was to perhaps make watches for his own label, which he simply called de Ries, at which point he thought of his nephew working at Omega.

Bannwart, ready to make a move toward independence and understanding that at the time practically the only possibility for getting it was to have a relative in the business, was amenable. Ries' personal invitation, extended at the end of the year 1954 and agreed to the following day, provided an opportunity to do something new. Bannwart was determined not to continue Ries' current trajectory or to approach Omega's existing clients.

Although Ries incredulously objected to the latter, he trusted in Bannwart's creativity and partnered with him, bringing his own wife Juliette—personal accounts reveal she was a very strong personality who believed in Bannwart's abilities—on board as well. Ries, Bannwart & Company was established on July 1, 1955. The trio had all of 50,000 Swiss francs and a credit card to their names; the company comprised a total of six individuals.

"My father always liked a triumvirate, because he said a chair on three legs is always steady."

Bannwart's third maxim was that his new company was going to make expensive, rare, limited-quantity watches. "I felt incapable of doing anything else," he later wrote. Though Gaston Ries was not exactly convinced that Bannwart's ideal was going to lead them to success, he went along.

They needed a new name to replace Ries, Bannwart & Co., though. And while they searched for it, they developed a logo: a golden key, which stood for *"la clef du temps parfait"* ("the key to perfect time"). This tagline would later develop into "the key to beautiful time" and even "the key to success."

Corum's first logo: the key to perfect time.

The actual brand name came along quite naturally shortly thereafter as a Latin word caught the partners' attention: *quorum*. The original meaning of this word is the minimum number of members of a body present in order to conduct business. The three partners decided to modify the spelling, thereby simplifying it and making it more pronounceable. "My father always liked a triumvirate, because he said a chair on three legs is always steady," his son, Jean-René Bannwart, remembers today. "With three people you always have one person against two others, and when one person can influence one of the two others, you usually win. If he cannot, the first idea was certainly not the best one."

René Bannwart, naturally, was the creative force behind Corum. And since he had developed his own artistic style at two companies of historic proportions, the ensuing watches were of course unlike much else that Switzerland had ever seen. Learning well from his time at Patek Philippe and breaking with the restrictions he had experienced at Omega during a time of global hardship, one of Bannwart's main pillars was the strict use of only premium case metals, which included 18-karat gold and the finest stainless steel, as opposed to the brass and platings in prevalent use in the watch industry at the time. "I consider plated gold to be a metal much more ordinary than a stainless steel that has been finished, polished, and ground," he told *Europa Star* in 1963. This in itself differentiated Bannwart's creations from those of the competition. Another rule he instituted in order to create products with his own stamp was to abstain from examining what other companies were releasing. "You can never completely avoid all influence from existing forms," he said in the same 1963 interview. "I think the secret is to follow your own path, to imagine new things that cheekily derive from one's own *esprit* in order to create truly novel models. You need to be at the prow of a ship and not in its wake."

"The secret is to follow your own path, to imagine new things that cheekily derive from one's own *esprit* in order to create truly novel models."

After releasing approximately 200 watches bearing the brand name Ries, Bannwart & Co., the new Corum introduced its first watch designed by Bannwart on the market in 1956, and in this year the small brand also made its premier appearance at the Basel Fair.

Gaston Ries passed away in 1958 and was replaced by his daughter and René's cousin Simone Ries. Together the two Ries women and René Bannwart turned the business into something extraordinary, something known for out-of-the-ordinary creations that were fashion statements in addition to being the finest examples of Swiss watchmaking.

The division of labor within the newly founded brand then located at 125 Rue Jardinière in La Chaux-de-Fonds was clear from the start: Bannwart was in charge of creation and even sales—not unlike his job at Omega—and the Ries women were in charge of organizational tasks such as purchasing and quality control and back-office duties like bookkeeping and invoicing. While he had been alive, Gaston had been the head of production, overseeing the other three employees, who were also watchmakers. René took this over thereafter. He and the two Rieses formed a three-legged stool whose benefits became apparent as Corum began to get underway as a brand; within five months the first little collection was ready.

As Bannwart began to visit Swiss retailers, he was often told that the watches were beautiful, but they would not sell for lack of a well-known brand name on the dial. This did not disturb Juliette Ries; everyone started somewhere, she thought. She came up with an astoundingly good idea: start sales outside of Switzerland, then come back with references in hand. So Bannwart began his journey in Paris, specifically with the most prestigious of all possible retail institutions: Cartier and Van Cleef & Arpels. These jewelers understood the value of creativity and partnered with Corum, resulting in dials on timepieces that at times read "Cartier by Corum" or "Van Cleef & Arpels by Corum." In fact, the first Corum watch sold was by Cartier.

Left: 125 Rue Jardinière in La Chaux-de-Fonds in 1955

Right: The Corum factory prior to 1995

From there, Bannwart went to Germany—historically an important market for mechanical watches with a large quantity of enthusiasts and consumers of such pieces. As predicted, retailers here were amply impressed by Corum's list of sales points and quickly got on board. Germany soon became Corum's biggest market by volume and remains an important one to this day, with the United States not far behind.

It was also Juliette Ries who urged that the company begin attending the Basel Mustermesse, as Baselworld was known in 1956, despite the horrendous cost to do so. "We have to do it," she simply stated.

And they never looked back.

Corum once boasted the world's longest wooden watchmaker's bench, which was found in the 1995 factory addition.

Above and left:
Assembling Caliber 13,
found in the Corum
Golden Bridge

A selection of pre-2000 Golden Bridge models; note the diverse designs thanks to Jean-Claude Schwab and René Bannwart.

CORUM
CORUM

A CHANCE ENCOUNTER

René's extroverted son, Jean-René, joined the company in 1966, bringing with him twenty years of experience in jewelry design. He also brought a love of sailing and an understanding of yachting sponsorships, which served Corum well in terms of marketing its first and biggest pillar collection, the Admiral's Cup. By 1970, there were twenty-five people working at Corum.

One of the employees who became integral in 1970 was Claude Roulet, who joined Corum that same year and who would be destined to remain at the La Chaux-de-Fonds brand until his retirement in 2012. Even today, he comes to the manufacture clad in jeans and comfortable shoes to work in the archives a couple times a month, thereby ensuring that his life's work remains ordered and preserved for posterity.

Jean-Pierre Moesch, general manager of Corum from 1958 to 2002

Claude Roulet received his own Golden Bridge in 1990 for twenty years of "precious collaboration."

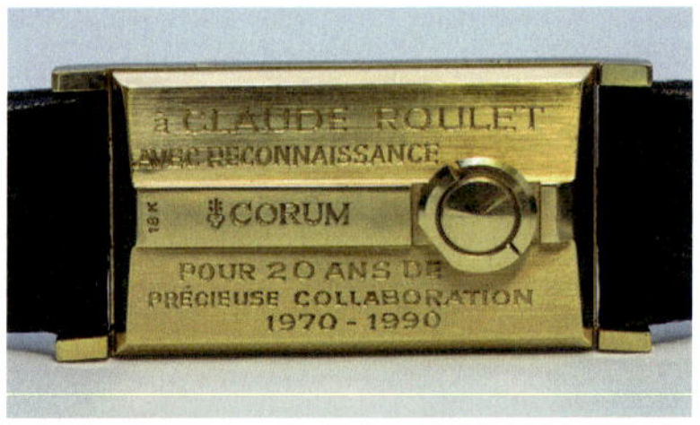

The back of Claude Roulet's Golden Bridge illustrates the interesting crown on the back of the case.

Another instrumental employee was Jean-Pierre Moesch, who joined Corum as the general manager in 1958 and remained there until his retirement in 2002. Today, Roulet describes Moesch as an "anchor," a crucial link in the chain of employees who made Corum what it is today. It was Moesch who maintained the close relationships to agents and clients spanning the globe.

Nothing escapes Roulet's memory, thanks in great part to copious notes and photographic elements arranged with so much precision that nothing could ever be forgotten. In 1970, Roulet was entrusted with the important task of stimulating media interest by means of advertising. Public relations, of course, did not yet exist as such. However, this form of supporting sales was so important to the company that Roulet was tasked with creating a special department just to coordinate the company's advertising, sponsoring activities, and the bit of what could be called PR that it did engage in. Today, we might call this the communications department. But in those days, there was no blueprint.

In the late 1970s, René Bannwart had a minor operation and was in the hospital for a few days. This in itself would not be worth mentioning except that in the bed right next to him was a man named Jean-Claude Schwab, who had also had a minor operation of some sort. The two did not know each other, but rapidly became acquainted as both were in the watch and jewelry industry: Schwab was a freelance designer who, at the time, was working for brands like Delaneau and had helped create a wristwatch made of wood. This piqued the Bannwarts' interest, as they were already working with stone and other unusual materials on their watches.

Schwab joined Corum and would become important not only to the story of the creation of the Golden Bridge, but also to Corum in general. He first improved some of the designs for existing products, including coming up with new dials, at the same time providing them a fresh lease on life in jewelers' showcases, but almost more important, he was the man translating Bannwart's ideas into tangible products.

Corum did not have a "creation department"; Schwab changed that and, once again, Corum was back to a triumvirate.

Jean-René Bannwart remembers that creations at Corum were generally the product of something akin to teamwork. "We were like a team of designers," he reminisces before going on to recall that he shared an office with his father for a decade, and they sat with their desks facing each other so that when one was away, the other could pick up the projects. "When my father was on a trip, and he had started with a design, I continued it. And when I was preparing something, talking with case makers, dial makers, etc., and I was on a trip, my father continued it. And when I came back, or he came back, we would look at what the other did in the meantime . . . and I would say, 'Wow, that's not what I was going to do, but it's not bad either'!"

Schwab found himself in the middle of all that, with father and son giving him ideas and Schwab making them viable, which included working with the case maker and dial maker. At that time, in a surprising difference from Omega of the late 1940s, Corum did not have a "creation department." The arrival of Schwab changed that.

And once again, Corum was back to being a triumvirate. Jean-René Bannwart took over the president's seat in 1987 when René was seventy-two years old. "My father didn't officially retire on a specific day," he remembers. "He slowly started to come less regularly to the office until November of 1992, when my parents moved to Lonay, near Morges. Even after that he continued to come to Corum in La Chaux-de-Fonds once a week for one or two days, spending a night in our house. That was a very nice period."

René Bannwart passed away at the age of ninety-four in January of 2010, just as Corum was marking its 55th anniversary.

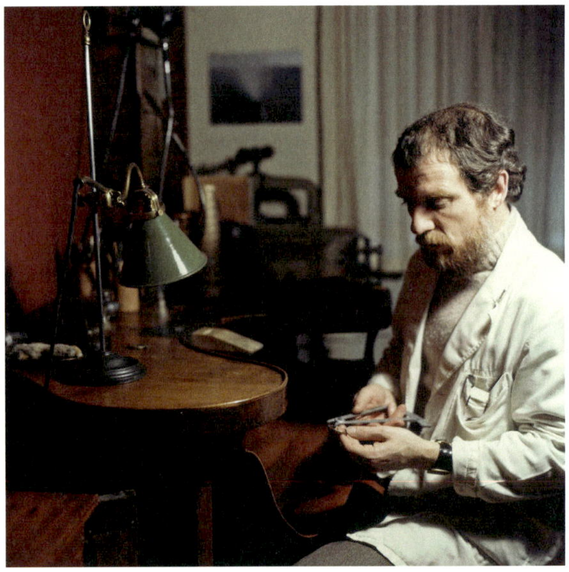

From top to bottom:
Jean-Claude Schwab, the first dedicated designer at Corum, in his workshop in the 1980s

Specialist watchmaker Gérard Aubry mills the 18-karat gold bridge of the Golden Bridge movement in the 1980s.

Serge Steudler, technical director and precision watchmaker, made the Golden Bridge movement fit for serial production in the early 1980s.

Jean-Claude Schwab hard at work designing variations of the Golden Bridge

BRIDGING THE MANUFACTURE

By 1991, it was apparent that the manufacture's edifice was too small. Jean-René Bannwart and Claude Roulet decided to take the opportunity to turn the building addition into something very special that would encourage young, artistic talents and also create some PR on its own.

Thus, Corum launched a competition to find the right architectural idea among freshly graduated young Swiss architects. This history-making move (this was the first time a private business in Switzerland had ever staged such a national competition) was received with an impressive amount of enthusiasm, and applicants sent Corum 116 accepted entries. The jury was made up of four architects—one of whom was Max Bill—a head of an arts school, and two representatives from Corum: Jean-René Bannwart and Serge Steudler, at the time technical director of the manufacture.

Seven of the 116 entries won prizes, while the grand prize winners, Margrit Althammer and René Hochuli from the ETHZ in Zurich, received the contract to design the new building, which opened its doors on June 16, 1995, just in time for the 40th anniversary of the company.

Left: jury members from the architecture competition including Max Bill (front row, right) Vincent Mangeat (front row, left) and Serge Steudler, Corum's technical director (far left, back row)

Right: Corum president Jean-René Bannwart (far left) and Vincent Mangeat (president of the architectural jury) with winners Margrit Althammer and René Hochuli

A highly modern structure characterized by a sober, yet transparent, austerity in concrete and glass, it soon contained something for the record books: the longest wooden watch bench in the world. But the reason that this building is included here is that the structure—likely the reason it won—was inspired by the shape of the Golden Bridge. The use of glass and steel even gives the building the same transparent look from the outside.

Top: The new Corum factory inaugurated in 1995

Above: a look inside the addition to Corum's factory today; note the showcased historical pieces on the top level and the clock by independent clock maker Jean Kazès hanging on the wall.

Watchmaker Roland Chapatte assembles the Golden Bridge in the 1980s.

RECOGNIZING THE ARTIST

In 1993, the Musée International d'Horlogerie in La Chaux-de-Fonds (MIH) established the Gaïa Award to honor the memory of one of the earliest patrons of the museum, Maurice Ditisheim. In sharp contrast to the Grand Prix d'Horlogerie de Genève, which can be seen more as the Academy Awards or Oscars, the Gaïa has often been called the Nobel Prize of the watch industry.

It is telling that in 1996 Vincent Calabrese (see chapter 2) was given the Gaïa Award for artisanal creation in 1996, and René Bannwart, the same award in 2000. Among the eighteen Gaïa Award laureates in this category since 1993 (awards are also given in the areas of historical research and spirit of enterprise, and not every category is awarded every year), Bannwart is the only corporate winner ever and the only winner who was not a watchmaker. Every other laureate is either an independent or a nearly independent watchmaker. Other recipients include technical watch industry luminaries such as George Daniels, Philippe Dufour, François-Paul Journe, and Michel Parmigiani as well as several of Calabrese's AHCI colleagues.

The museum describes the artisanal creation category in these words: "It is without doubt the desire of creative and audacious watchmakers operating within relative anonymity, whose names were discreetly associated with great companies, that pushed the initiators of the prize to want to distinguish them above all. Rich, inventive and tenacious personalities, just ten years ago their work was never recognized by a non-specialized public not passionately involved in horology like it is today. It pleases us to believe that the Gaïa Award has at least modestly permitted people to discover the work of some of these genius artisans."

Left: René Bannwart (center) with Simone Bédat (left) and Kathleen Pritchard (right) receiving the 2000 Gaïa Award

Right: Vincent Calabrese (center) with Günter Blümlein (left) and Jean-Luc Mayaud (right) receiving the 1996 Gaïa Award

II

VINCENT CALABRESE'S LINEAR INVENTION

VINCENT CALABRESE'S LINEAR INVENTION

Vincent Calabrese is a self-taught watchmaker born in 1944 who has earned a degree of fame in the watch industry as the cofounder of the AHCI (Académie Horlogère des Créateurs Indépendants). This group of independent creators founded in 1985 counts some of the most creative watchmakers in the entire world among its members.

Something of a philosopher, Calabrese's handmade designs challenge the traditions of mechanical engineering, and today we also know that they tend to defy usual horological practices in their originality and diversified inspiration.

Even as a young man in the 1970s, Calabrese lived his nonconformist attitude. At the age of thirteen in his hometown of Naples, Italy, he began working for a local watchmaker at the suggestion of his mother, replacing an apprentice for a short time. In that brief period, after having had to plunder his family's scant savings to buy his own tools, his natural ability allowed him to learn to repair watches, which he then began doing for a living. He moved to Switzerland in 1961 at the age of seventeen to learn more about horology as well as to escape the confines of a conservative society, which required army conscription and in which he did not necessarily feel he belonged. Calabrese felt a great sense of liberty upon arriving in the snowy Swiss Jura mountains.

He moved to Switzerland in 1961 at the age of seventeen to learn more about horology as well as to escape the confines of a conservative society.

He worked for a few large brands, learning much about the making of watches as well as how to fend off the prejudice he encountered for not being a Swiss watchmaker, and finally settled in as shop manager for a retail location in Crans-sur-Sierre, a small place located in Crans-Montana, a well-known ski resort in the heart of the Swiss Alps. It goes without saying that it is a chic place: famous personalities such as Roger Moore and Sophia Loren own property there, and Calabrese counted celebrities such as Gina Lollobrigida and Catherine Deneuve among his cherished clients.

The year was 1972, and despite his luxurious surroundings and enjoyment of the relaxed atmosphere—a far cry from the chaotic port city where he grew up as well as the pervasive discrimination of the Swiss watch factory workers—Calabrese was feeling quite

Top: Vincent Calabrese in 1985 in his workshop

Bottom: Naples, Italy

Vincent Calabrese's original Golden Bridge prototype movement (photo Ian Skellern)

A Spatiale watch by Vincent Calabrese with the movement's bridge formed as the letter C; the highest form of personalization dissatisfied. The store he worked in had every luxury brand available. But many of his important clients asked him for personalized watches when they came in. His desire to please them gave him the idea of crafting a movement that was personal, and therefore easily customizable, but also structural, airy, dimensional—spatiale in French. "This was the means of being able to personalize a watch to the maximum," Calabrese remembers. As he was compelled to make his movement as an artisanal piece, every time he executed it completely by hand it was new and fresh.

"I ceased to 'seem' and began to 'be.'"

But it was also at this moment that Calabrese ceased to be just the self-taught rhabilleur watchmaker and began to become something else: a designer, a creator, a thinker—and a manufacturer who needed to purchase tools, machines, and materials in order to complete his orders.

"I ceased to 'seem' and began to 'be'," he says now. Calabrese cites this as a turning point in both his life and his artisanal education, and he decided to jump in with both feet and learn all the crafts he would need to be able to make an entire watch. And then one further event provided Calabrese with the impulse to start down the path of his philosophical journey. He had a rich Parisian client who owned a 200-year-old Breguet watch. It had been dropped and was run over by a car. There was damage, but it was reparable. However, at the time there were few watchmakers with the skills to fix it, so the estimate for the repair was quite high. The client chose to only rehab the case and not the movement, arguing that one can only see the case anyway. "This can't be possible," Calabrese thought to himself, "Breguet is the most esteemed watchmaker in history."

This experience solidified Calabrese's decision to make a watch in which nothing was visible but the movement—with no dial to hide it, no case to distort it, and an in-line movement architecture that made it relatively easy to incorporate shapes and letters. It took him two years of head-scratching, immersive work, and when he was done in 1977, he took his single-bridge movement to Geneva's Invention Salon, where it won the gold medal. "After that, I had many requests from watch companies," he remembers.

But before he had even arrived at the exhibition, he had used a couple of his own contacts to set up visits at two companies with whom he thought he might like to work and to which he had thought to sell the movement idea. One was Patek Philippe. "I went to see Philippe Stern, who had just taken over from his father Henry, and showed him my prototype. He was very interested and thought about it for a month or two, and then he said no, that he would love to, but it wasn't like the things they were already doing."

CORUM AS AN INDUSTRIAL PARTNER

Of all the companies Calabrese was in contact with the second one he chose was Corum. One must remember that this was not yet the era of the independent watchmaker, a concept as yet unknown for all intents and purposes. "Corum was a dynamic company that made many things," he fondly remembers before explaining how well he clicked with Corum's owner René Bannwart right from the get-go. Others inside modern-day Corum who were there at the time have remarked that they recall it as a meeting of two like minds or a dialogue between two artists. Years later Calabrese continues to describe Bannwart as an exceptional person. "The curator of the Musée International d'Horlogerie at La Chaux-de-Fonds, André Curtit [with whom Calabrese was quite friendly at the time], had a very good relationship with Bannwart. He called him [on Calabrese's behalf] at eight in the morning. By nine I was sitting with him. And by noon I had signed the contract. The Golden Bridge was born."

Making individual pieces of his Spatiale movement by hand was one thing. But producing it for use in a serial timepiece was quite another. To complicate matters, both Calabrese and Bannwart wanted the movement to be produced in 18-karat gold, a material that is very precious, but very soft and relatively difficult to manage for stable horological use.

"He called him at eight in the morning. By nine I was sitting with him. And by noon I had signed the contract. The Golden Bridge was born."

"The reason for this choice was simple," Calabrese remembers. "It was to be an exceptional movement. I did not want it to be confused with any other mechanism. Up to then, there had never been a wristwatch movement in 18-karat gold."

Using a gold that was so pure (18-karat gold is 75 percent pure gold) did not end up damaging the integrity of the Golden Bridge's large visible bridge, which more or less hid the wonder of the interplay of wheels, springs, and pinions, though technically speaking, it was perhaps a bit less stable than a straight-line bridge made from a more common horological material or a gold plate or bridge with more girth, as would be found in a normal round movement. But Calabrese, who had by then had training in

Opposite: Vincent Calabrese's own Régulus tourbillon, which adds a flying tourbillon escapement and regulator display to a linear movement

VICA
750

CORUM
E 274/350
SWISS MADE - WATER-RESISTANT 30 M

the art of jewelry-making, knew the properties of gold well and was unconcerned. "It's a small detail that enriched the object," Calabrese opines. The name of the watch obviously comes right from this technical detail: it's a bridge, it's in gold. Voilà: the Golden Bridge!

Serially producing the movement proved to be too much for Calabrese alone, and Corum moved the operation to its own facilities. It is important to note two things about manufacturing in 1980 that we take for granted today:

1) CNC machinery and other large tools allowing the precise serial production of components did not exist yet, especially not commonly in factories like today. Large quantities of components were stamped, which reduced their precision as compared with milling, and they needed a great deal of finishing by hand afterward. Fortuitously, the Golden Bridge and other Corum collection staples were not necessarily sold in extensive quantities. The components were traditionally produced by suppliers, who delivered them to the factory, where Corum's own watchmakers assembled, adjusted, and finished them.

It's a bridge, it's in gold. Voilà: the Golden Bridge!

2) The term *manufacture* was not yet significant. Most companies produced watches using the traditional division of labor prevalent in Switzerland, which made for a better product in the end, as the individual components were crafted by specialists. Making the Golden Bridge movement inside Corum's own four walls, which was possible thanks to its minimalist nature, did elevate the brand to "manufacture" status, though.

Naturally, this way of working precipitated a few changes to the movement, as it was no longer one single watchmaker doing everything; the movement design had to be adjusted for serial production as well, but this term should be taken lightly. From 1980 on, the Golden Bridge experienced an annual production in the low hundreds—meaning that it is still rare enough to remain an exceptional eye-catcher.

Opposite top: Vincent Calabrese's personal Golden Bridge with the bridge in 18-karat white gold

Opposite bottom: The back of Vincent Calabrese's personal Golden Bridge

THE INDEPENDENT WATCHMAKER

In modern times, the term "independent watchmaker" signifies a lone creator without brand affiliation. This era generally began with the work of George Daniels in Britain a number of years before Calabrese in Switzerland began to form the loose grouping of independent creators called the AHCI. It is interesting to note that Daniels, too, was self-taught, but it is also important to remember that no creator ever reached the completeness of Daniels' abilities; he was able to manufacture an entire watch by himself aside from the crystal, though each of his creations was never executed more than once.

The Golden Bridge marked the beginning of an era in which the artisan would step forward out of the shadows.

One thing all independent watchmakers have always had in common since that time is a constant struggle to survive in a tough business environment. With a very few lucky exceptions, most of them live and work right on the edge of viability; in other words, one watch sold represents the ticket to surviving a few more months and having the means to create the next watch.

With the idea and implementation of the Spatiale with its customizable in-line movement, Calabrese jumped into a new world of self-fulfillment and creation that depended on his ability not only to successfully produce unique watches, but also to successfully sell them. This in itself is generally the key element making or breaking the career of an independent watchmaker.

When Calabrese won the gold medal at Geneva's Invention Salon and a few articles followed in the media, he expected orders to arrive at his doorstep of their own accord. But, disappointingly, this did not happen. "Why?" Calabrese rhetorically asks. "First off, because I have an Italian name. Secondly, because I was an artisan, and people were not used to buying directly from artisans." This reaction allowed a simple reality to dawn on him: "The artisans were generally concealed/obscured by the brands; they were never able to sign their own pieces." And this was the revelation that allowed him to create two things: first, the Golden Bridge, and second the AHCI (which Daniels ended up becoming part of as an honorary member).

"This one watch formed the base for everything, for my entire life. It was no longer the money or the success that interested me, but rather me and my life. I was allowed to live my life, not the life of others; I was allowed to create my AHCI academy and all the things that came after."

This one wristwatch was more than just a timepiece; it was a message. For Calabrese the message was that both the work and name of the watchmaker would no longer be hidden. For Corum the message was that mechanics and art can coexist to make a wearable object of fashionable beauty.

And the interesting thing was that this was fully accepted by the consumers, the retailers, the critics, and the artisans. The Golden Bridge was part of something new, something progressive, something forward-thinking that would mark a turning point in a subculture of a world that increasingly focused on the soulless perfection of quartz. The Golden Bridge, with all of its imperfect mechanics, its collaborative nature, its inspired genesis, marked the beginning of an era in which the artisan would step forward out of the shadows, if ever so lightly and ever so tenuously.

"This watch communicates from the inside out; it is a creation that was created from the gut, not the head. And the head had to realize that the gut was dictating everything," Calabrese remembers.

Left: Vincent Calabrese in 1989 with his unusual tourbillon table clock

Right: Vincent Calabrese today in his workshop near Lausanne (photo Ian Skellern)

This one watch allowed Calabrese to win the biggest battle that every single independent watchmaker engages in. It allowed him to continue his craft, to "win" in a sense. And Calabrese himself says that his life was definitively divided into before and after the Golden Bridge. "Every day I battled for my ideas of equality, liberty, the dream of the academy, to show watchmakers all over the world they can do things, create watches. And I succeeded. For thirty years I have succeeded. How many are so lucky?"

PATENTS

The Golden Bridge was awarded seven patents in total: one, which Calabrese authored, was for the winding and setting device, which allowed the watch to be wound and set from the back, thus not disturbing the watch design with a visible crown. Another patent was awarded for the sapphire crystal case, and five more were issued for the design of various models including various wrist-watch types, the pendant watch, and the table clock achieved in collaboration with Baccarat.

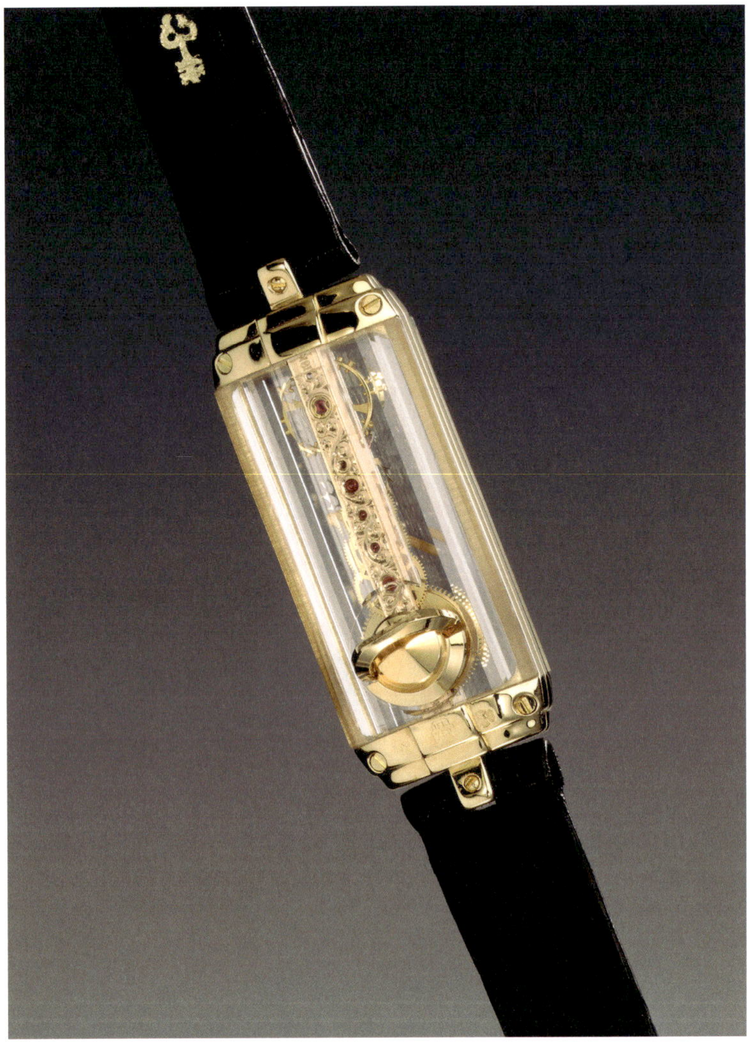

The original Golden Bridge with crown for winding and setting "hidden" on the back

DEVICE FOR WINDING AND SETTING A TIMEPIECE

PATENT NUMBER 4362395

ABSTRACT A winding and setting arrangement for a timepiece comprises a two-part stem extending through the hollow shaft of the spring drum. The outer part of the stem is affixed at its outer end to a crown and at its inner end has Breguet teeth which, in an inner position of the stem, are engageable with Breguet teeth of a connecting member, which is non-rotatable but axially movable in a cavity in the hollow spring drum shaft. A spring urges the connecting member outwardly against a shoulder in the shaft. The inner part of the stem is coupled with the outer part by a square section so that the two parts rotate together but are axially movable relative to one another. The inner stem part passes freely through the connecting member and, inwardly of the spring drum, carries a pinion which is engageable with a setting wheel in an outer position of the stem. A spring urges the inner stem part outwardly.

FILED October 10, 1980

ISSUED December 7, 1982

ASSIGNEE Corum, Ries, Bannwart & Co.

INVENTOR Vincent Calabrese

CORUM
CORUM

III

A TRANSPARENT MASTERPIECE

A TRANSPARENT MASTERPIECE

The Corum Golden Bridge was a watch to mark history by, both in its design and technology. In the previous chapter, we learned how Vincent Calabrese had the idea of a naked, visible movement that left no technical trick to the imagination, while René Bannwart took the idea to the next level by turning it into a collection that could be serially reproduced (more on that evolution in chapter 5).

The third and final piece of the puzzle was making the transparent case as envisioned by Bannwart, Calabrese, and Jean-Claude Schwab into a tangible reality. This was nothing short of a technical marvel in 1980, when sapphire crystal had not yet been established as the material of choice to cover a watch dial. In 1980, manufacturers were as a rule still using either Plexiglas or mineral glass to cover dials; synthetic corundum—though in wide horological use in the form of synthetic rubies as bearing jewels within the movement—was still a relative rarity as a dial covering.

Bannwart never forgot that a case's dimensions needed to naturally correspond to the anatomy of the human being wearing it. Round cases had generally been derived from the shape of the movement within and not the shape of the arm. He also felt that women could indeed wear larger wristwatches if they were shaped correctly. This is the reason why only one unisex Golden Bridge model was created in the beginning.

"We didn't want gold on the side, we wanted complete sapphire."

A solution had to be found as to how to craft a full case made of sapphire crystal. "The problem," Jean-René Bannwart remembers, "is that at that time sapphire crystal was not easily worked like today. It was quite difficult. And we didn't want gold on the side, we wanted complete sapphire."

The obvious characteristics of sapphire crystal made it the impeccable choice for use in the Golden Bridge concept: perfect transparency from every angle; great durability; permanent luster exuding luxury and brilliance thanks to its high refraction index, not dissimilar to diamond; and its extreme resistance to scratching.

The very first version of the Golden Bridge comprises two hand-faceted sapphire crystals held together by four golden screws and two bases of yellow gold. The model is completely elegant in its transparency, and there is no gold disturbing the intense luster of the sapphire crystal.

Top: The first serial piece of the Golden Bridge numbered 001, which today is on display at the Musée International d'Horlogerie (MIH) in La Chaux-de-Fonds

Above: The first serial piece of the Golden Bridge numbered 001 (left) next to Claude Roulet's personal watch

MAKING SYNTHETIC SAPPHIRE CRYSTAL

Synthetic sapphire crystal came into use in the late 1960s, and applying this material on timepieces had three distinct advantages over the acrylic and mineral glass types of crystal that had mainly been used until then. For one thing, synthetic sapphire is hard (9 on the Mohs scale and just about 3,000 on the Vickers scale), and therefore it can only be scratched, or even worked, using a diamond. For another, synthetic sapphire is heavy, lending any watch using it an even more luxurious feel. And finally, synthetic sapphire is still a precious stone, albeit artificial, making the watch more valuable in its entirety. The only drawback to synthetic sapphire crystal was that it was prohibitively expensive to source as needed, and this was due to the difficulty that suppliers had in working it, especially when unusual shapes were needed.

The story of the synthetic sapphire crystal began in the year 1902 when Professor Auguste Victor Louis Verneuil announced that he had discovered how to duplicate the chemical composition and crystal structure of a natural ruby; two years later he published a paper explaining the process for producing colorless synthetic corundum, the raw material of today's sapphire crystals. Verneuil's original process has remained basically unchanged until today.

What nature needs about 100,000 years to create is completed here in about fifteen hours.

This process begins with powdered aluminum oxide (Al_2O_3) that is flame-fused by being introduced into a blowpipe invented by Verneuil that is heated to a temperature of 2,050°C by mixing hydrogen and oxygen. A mechanically activated hammer at the top of the blowpipe taps the reservoir of alumina powder, causing drops of it to fall regularly onto a prepared "seed crystal." What nature needs about 100,000 years to create is completed here in about fifteen hours: the corundum—or crystallized alumina—has formed on the rod and now looks similar to a stalagmite. This "stalagmite" of corundum, which is called a *boule* in proper horological jargon, is then refired at a temperature of 1,800°C to stabilize the crystal. At this point, the job of the crystal grower is done, and the raw corundum boules are sold to watch crystal manufacturers, who pay by the carat and according to the diameters they need.

Once you've understood this process, it is not hard to spot that unusual shapes such as that of the original Golden Bridge case cause difficulties in every step of the supply chain, making them expensive and rare. There is also one other factor that makes sapphire crystal very expensive: any imperfections in the material are only visible after countless hours of polishing and working, and generally about 15 percent of all corundum is unusable due to inclusions.

Sawing to the approximate shape using tools sporting diamond tips takes up to eight hours, after which the blanks are ground to the precise dimensions within a tolerance of 200ths of a millimeter. During these two processes alone, more than half of the corundum material is lost in the form of dust. Then the thickness is evened out, the exterior shape formed, and the crystal beveled, faceted, and recessed. It will then be domed, polished, and cleaned, with minute quality controls performed after every single step. Only after about twenty operations have been completed are any imperfections visible in the material.

One other factor makes sapphire crystal very expensive: any imperfections in the material are only visible after countless hours of polishing and working.

The Swatch Group is the only Swiss company that has the ability to produce and work synthetic sapphire crystal from start to finish thanks to the many mergers that have formed Comadur SA, which is headquartered in Le Locle. Boules are created at the factory in Courtepin and at Seitz in Les Brenets.

Seitz, founded in 1927, is the company that manufactured the original crystals for the Golden Bridge. Seitz began its career by manufacturing rubies for use in watch movements. By 1956, it had extended its scope to related technologies so far that by 1980 Seitz was no longer manufacturing jewels, though it did continue working with corundum and other hard materials. In 1968, Pierres Holding was formed, uniting about twelve companies working in this same field. Pierres Holding was eventually assimilated into what is today the Swatch Group to form Comadur.

Seitz was so proud of its achievement that it included photos of the Golden Bridge and its unique crystal in its company brochure. Additionally—and uniquely in the watch industry—Seitz took part in the official press conference held to introduce the Golden Bridge.

PRESS CONFERENCE AT THE MIH

On September 25, 1980, Corum and the maker of the sapphire crystal case housing the Golden Bridge movement, Seitz AG, staged a press conference at the Musée International d'Horlogerie (MIH) in La Chaux-de-Fonds. René Bannwart spoke on behalf of Corum and managing director Jean Grédy, who passed away in 2009 at the age of seventy-six, represented Seitz.

"If this model arouses admiration and envy in all who see it, the reason is the sheer beauty of the watch, demonstrating the watchmaker's craft at its very best."

While the press conference focused on the visibility of the new movement and the technical marvel of the sapphire crystal, it is very important to note that Bannwart did not forget to mention Vincent Calabrese as the movement's inventor, as a transcript of his statement verifies. This is significant, particularly because it is so unusual in the watch industry to give mechanical credit where credit is due where outside assistance is concerned. This was, after all, the same era (though a couple years later) in which today's most famous independent watchmaker, Philippe Dufour, was so dissatisfied with one of the grandest brand names in the entire watch industry not even allowing him to set foot in its Basel booth after he had created a grand complication for it that he struck out on his own as an independent. For Bannwart to not only honor Calabrese, but to publicly hold him in the highest regard at the press conference in 1980, is nothing short of spectacular. It remains a distinction that is rarely seen even today. "[It] immediately struck the dedicated watch craftsman in us as being an innovation with inspiring potential. From the very beginning, we admired the pure beauty of this movement and realized that we could dress it superbly," Bannwart said during the press conference.

"Together with Seitz, we have succeeded in creating a richly faceted sapphire case, completing it with gold attachments to which the leather bracelet is affixed. This was the most outstanding achievement in our Golden Bridge collection," Bannwart proudly declared.

Transparent cases have popped up occasionally throughout the history of horology, but until the Golden Bridge, these were generally made of rock crystal or mineral glass. In his statement, Bannwart explained that the Corum sapphire crystal case would not have been possible even two years prior thanks to the difficulty of the faceting, which was only achieved due to a new technique that Seitz had devised and by then had mastered.

For his part Grédy also seemed gratified, as a transcript of his remarks showed. "Seitz is the first company in the world to make a watch case from sapphire crystal and bring it to market. The idea of encasing a movement in transparent material is not new. Here at the museum we can point to a timepiece from the seventeenth century, whose case is made of rock crystal. Despite this, even two years ago it would not have been possible to realize a case like the one we are presenting today."

Left: René Bannwart (left) and Jean Grédy of Seitz SA present the Golden Bridge at a press conference at the MIH on September 25, 1980.

Right: René Bannwart and Jean Grédy present Pierre Imhof, president of the Musée International d'Horlogerie in La Chaux-de-Fonds, with number 001 of the Golden Bridge.

Foreshadowing the passion that would be aroused in the hearts of collectors, aficionados, and lovers of fine mechanics by not only this model to become a fervor once again directed toward all fine mechanical watches, Bannwart closed the press conference with this statement: "If this model arouses admiration and envy in all who see it, the reason is not merely the high precision of the movement, which will henceforth be the domain of electronics, but because of the sheer beauty of the watch, demonstrating the watchmaker's craft at its very best."

At the end of the press conference, from which excerpts were broadcast on Swiss television and radio over the ensuing few weeks in addition to being written about in approximately 200 magazines, he offered the very first serial example of the watch, which bears the number 001, to the museum. The Golden Bridge began arriving in Corum's retail locations within the same year.

This original serial Golden Bridge still on display at the MIH is unlike any other Golden Bridge anywhere else in the world today. As the goal of both the Bannwarts and Schwab was complete transparency, the first solution they attempted was to glue the two sapphire crystals together and to the gold brackets for securing the strap. Though it remains common practice to glue certain elements of a watch together—even the crystal—this particular solution did not work out as the trio of Corum creatives had hoped. "This example trapped the movement inside the case; there's no gold reflection anywhere," Jean-René Bannwart reminisces regarding this first attempt at Corum's ideal of absolute beauty. "But, of course, you couldn't see through it on the side anyway because [the crystal] was too thick."

The younger Bannwart also recalls that the case maker guaranteed that the glue would hold for centuries. But, unfortunately, the first pieces were already coming back after three months. It would seem that the case manufacturer didn't take the heat of the spotlights in the retailers' showcases in conjunction with various skin care products into consideration. This meant that Schwab had to redesign the case to include a thin strip of gold along the side. This was easily accomplished by combining the glue with a little groove in the metal, much like the setting of a gemstone.

The final technical feat was the hole in the crystal for the crown on the back. It had to be drilled out using precision diamond-tipped tools and also made water-resistant with a gasket—all without compromising the absolute transparency and beauty of the material.

The design of today's Golden Bridge has evolved somewhat to accommodate a modern feel for sizes and trends. The models are no longer unisex, and the crystals are now manufactured in three sizes (Golden Bridge Mechanical, Miss Golden Bridge, and Golden Bridge Automatic). Though unusually shaped sapphire crystals are more familiar today than they were in 1980—including in Corum's own Bubble as designed by Severin Wunderman more than two decades later (see next chapter)—they are by no means ever off-the-rack products. "This item will never be manufactured in industrial amounts," Grédy remarked during the press conference. "However, it stands as a reference point, a benchmark for the state of professional skill."

The Golden Bridge remains the unique example of technical artistry René Bannwart had envisioned when he cofounded Corum in 1955.

Opposite: The original sapphire crystal of Corum's Golden Bridge

Period advertising

Advertisement from 1997

IV

THE MODERN ERA

THE MODERN ERA

Severin Wunderman was born in Brussels, Belgium, on November 19, 1938, but moved to Los Angeles, California as a teenager in the early 1950s to live with his sister, Bella, due to the death of his parents. Upon leaving school in 1956—but not graduating with a diploma, as he was more for practical business learning than book learning—he returned to Paris, which he has cited as his favorite city. He related a story to the magazine *LifeStyles* in 2002 in which he describes having gone on to work as a watchmaker apprentice in Belgium in 1958. "One day," he declared, "I had just received my salary (1,700 Belgian francs), and on the way home I passed an antiques store. I saw Jean Cocteau's drawing 'Les Enfants Terribles.' It cost 1,600 francs. I didn't hesitate for a moment; I succumbed to my passion and bought it."

Two years later, he had the opportunity to help out at the one-man company of a distant relative who made his living as an importer of wristwatches in Belgium. Following a heart attack, the man could no longer drive his own car, so he offered nineteen-year-old Severin the chance to work as his chauffeur on business trips. "I became his driver and carried his sample case," Wunderman remembered in an extensive interview with German watch magazine *ArmbandUhren* from 2005 as Corum turned fifty years old.

"I saw Jean Cocteau's drawing 'Les Enfants Terribles.' I didn't hesitate for a moment; I succumbed to my passion and bought it."

Wunderman explains that the business at that time was quite friendly and personal among the jewelers, importers, and watchmakers. For this reason, he was able to learn a lot very quickly. His relative passed away suddenly when Wunderman was twenty-four, and he took over the business. At first he was the importer for the Swiss brand Juvenia. Then he went to the French watch company Alexis Barthelay, where he learned about the art of private labeling, as Barthelay was behind certain models for high-end brands like Cartier, Van Cleef & Arpels, and Tiffany.

Opposite: The first versions of the modern Golden Bridge and Miss Golden Bridge

CORUM
CORUM

ANOTHER LIFE-ALTERING CHANCE ENCOUNTER

A chance encounter with Aldo Gucci changed the course of Wunderman's life completely: at a jeweler's shop he was visiting for Barthelay, the story goes that he picked up the receiver of a ringing telephone to find Gucci on the other end of the line looking for someone else. The man from the famous fashion house reacted in an annoyed manner, but Wunderman—a polyglot versed in several European languages—answered him in the same Florentine dialect. Gucci was yelling at him in and they ended up friends. "He became my mentor, my best friend, and he really taught me everything I needed to know about business," Wunderman remembered in an interview with The Purists in 2003 regarding his forty years in the watch business.

"I am 100 percent entrepreneur and don't mind taking risks."

The precise details of the story vary slightly in different media outlets, with *The House of Gucci* by Sara Gay Forden, a juicy tale of the soap opera-like Gucci family, including another version. This leads me to believe that Wunderman may have had his fun at the expense of the media at times and wasn't necessarily very interested in divulging too many personal facts. Or perhaps this was part of his special brand of creativity. Wunderman never did provide media representatives with much of an official bio or curriculum vitae; the narrative you are currently reading was put together using several different sources, including firsthand accounts, memories, and interviews.

The fact does remain, though, that Gucci became his best client when Wunderman suggested a licensing agreement after working with Aldo Gucci for two years. "I founded Severin Montres in 1972 in Switzerland and took up production," he told *ArmbandUhren*. "There was a price segment gap at the time between $150 and $395; I positioned watches with a world-famous name right there, including unusual models like ring watches. Gucci sold the largest number of units of all Swiss watches and was number two behind Rolex in value." In fact, he was producing about one million units per year, sold in more than 3,500 points of sale globally.

Wunderman, who paid the highest price of all licensing deals in the watch industry at 15 percent, had received carte blanche from his mentor and designed all the licensed watches within his own organization. "I am 100 percent entrepreneur and don't mind taking risks," Wunderman continued in the *ArmbandUhren* interview. "One of the reasons for my success is that I always swim against the current. For example, I was the first to put diamonds on stainless steel watches. At the time, everyone said that was a sin, but do you know one company today who doesn't put diamonds on steel? And that includes Patek Philippe."

Wunderman had been producing Gucci watches under the licensing agreement for twenty-three years, going from zero to 550 million francs' worth of turnover, when in 1997 Gucci decided to take full control of the watches made under its name. This left Wunderman room for another watch brand in his now-sizable stable of companies. The sale of Corum, made necessary due to irreversible economic woes, came just at the right time for him.

"The watch industry is something that enters the bloodstream like a drug."

Though Wunderman also owned other companies, he remained particularly fond of the watch business, and Corum was his smallest operation. "The watch industry is something that enters the bloodstream like a drug," he remarked when this was pointed out to him.

A passionate collector of both antiques and art, Wunderman was especially interested in the work of Jean Cocteau. A great supporter, by all accounts he had the largest collection of Cocteaus in the world and according to one source had amassed 2,000 individual pieces. He founded the Severin Wunderman Museum dedicated to Jean Cocteau in Orange County, California, which closed in 1995. He also became the most important contributor to the new museum in Menton, France, which opened its doors in 2010 to display 1,200 of Cocteau's works—all donated by Wunderman. His lifelong commitment to the artist brought him the respect and acknowledgment of Cocteau's home country, and on December 31, 2004, he was appointed a knight of the National Order of the Legion of Honor by the president of the French Republic.

Wunderman loved art, especially Cocteau, and some of his designs for Corum clearly exhibit the artist's influence and reflect the kind of classic elegance that runs through his work. Wunderman also dabbled in Hollywood, even marrying an actress before divorcing her four months later. When asked about his inspiration for designing in a March 2002 interview with journalist Marie-Clare Borle, he said, "I generally draw much of my inspiration from the works of Cocteau because he has an extremely linear style."

After narrowly escaping both the Holocaust (by hiding in a convent housing blind children) and cancer (after being given only one month to live) earlier in his life, and openly admitting a fascination with death (resulting in many watches' skull designs), he passed away on June 25, 2008, in Nice as the result of a severe stroke. In the last decade of his life, the cosmopolitan businessman and art lover, who had been married four times, had residences in Los Angeles, London, Nice, and La Chaux-de-Fonds.

CREATIVITY AS THE CONTINUATION OF THE BANNWART IDEALS

Wunderman was among the shrewdest businessmen in the watch industry's recent memory. To ensure his success at "polishing up" the dusty jewel that he called Corum, he brought most of his Gucci team intact to the new brand, a staff he called the "dream team" at the time. He also made sure Jean-René Bannwart, who still owned 10 percent of the company for another year, remained on the brand's board to protect its patrimony.

One outward expression of Wunderman's unquenchable creativity was the brand's Baselworld booths during his time at the helm. Unlike René Bannwart, who decorated the booth's windows, Wunderman decorated the entire booth in a new way every single year he was in charge. At Baselworld 2003, for example, the booth was done up to represent a modern cabinet of curiosities, blending art and nature with dreams and reality. On exhibition inside the booth, Wunderman had twelve works of art by renowned Swiss painter and sculptor Hans Erni (who is now 105 years old). These modern "vanitas" were chosen to remind Wunderman and all who saw them that there is no frontier between life and death and to match the curious objects in the display windows, which caused a great deal of chatter among visitors to the fair. Another year the whole booth housed paintings from his own art collection, and yet another year he decorated the outside of the booth completely in galuchat.

"He loved his people, he always took care of them, but it took a good year for [the Swiss employees] to understand that he was on their side."

Most important, Wunderman was human in a very real sense, and he genuinely loved his employees. Stacie Orloff, president of Corum USA for seven years starting in 2000 and a member of Wunderman's "dream team" for more than twenty years, remembers what it was like to work with him. "He loved his people, he always took care of them, but it took a good year for [the Swiss employees] to understand that he was on their side," she remarks on some of the difficulties with differences in mentality. "We were all together on 9/11 in a meeting at Corum in Switzerland . . . the World Trade Center tower went down, and then the next one, and then we were all huddled in the car in the parking lot because we didn't have satellite then. And then Severin flew us all to his house in France because we couldn't get home to the United States."

The Corum booth at Baselworld 2006 done up in 1920s-style art deco as expressed by walls fully covered in galuchat, black-and-bronze geometry and walnut parquet floors; the paintings in the booth windows were from Severin Wunderman's own art collection.

Wunderman remarked on the hard work put in by all in an interview published in March of 2001 with journalist Luc Giacobino when asked to describe his first year as Corum's head. "A great deal of hard work, exceptional results . . . with a fantastic team! Judge it on the facts: in 2000 we practically doubled our sales, which went from thirty to fifty-four million Swiss francs, transforming the losses of 1999 into profits. As for our production, it has leapt from 9,000 to 40,000 pieces. In fact, this performance was achieved in only six months, mainly in the course of the second half-year . . . which makes us very confident for the future."

"That was Severin's spirit; I think he was challenging and artistic, and maybe misunderstood by the Swiss, but by no means was he ever disrespectful to their heritage. He really embraced watchmaking for what it was. He was a collector, a true collector," says Orloff.

The Golden Bridge,
mid-2000s

Top: A Golden Bridge variation from the mid-2000s

Above: A fully diamond-set Golden Bridge in the refreshed case of the later 2000s

THE GOLDEN BRIDGE RELAUNCHES FOR FIFTY YEARS OF CORUM

Orloff, who comes from a jeweler's family, is convinced that it was the Golden Bridge that drew Wunderman to Corum in the first place. "It was one of his motivators for buying the company, and he was respectful of it. We were a design-led company, and that was the prize."

Wunderman seemed to agree, as he remarked in the 2001 interview: "Corum is without a doubt a gem. It was asking to be polished a little. I am personally devoted to it with all my strength, with all my heart. It won't be long until it shines all by itself."

A year later, he said the following in an interview with Borle: "Corum and I were the perfect match: I was attracted by this company, which had always done things differently from others. Some of its models have remained great classics: the Admiral's Cup, the Coin Watch, the Golden Bridge. Despite its youthfulness, it has already stored up an impressive legacy in its archives."

Oral narratives have revealed that Wunderman became a tad exasperated, however, when the Golden Bridges would come back to the factory for repairs. The mechanism was rather sensitive, and by the time he was heading up the brand, the in-house watchmakers able to fix the Golden Bridge movement had left. It became such an issue that Corum began to offer a new Coin Watch as part of an exchange for Golden Bridges that came back for repair. "He would only take this project on if he could modify the movement to make it reliable while maintaining the beautiful integrity of the Golden Bridge, the genius of what it had to offer," Orloff remembers.

"Corum and I were the perfect match: I was attracted by this company, which had always done things differently."

Additionally, Wunderman, whom Orloff recalls as never really wearing a watch, although she knows he collected outstanding examples from Rolex, Patek Philippe, and Corum, among others, was concerned about the unisex style of the original Golden Bridge. He felt there needed to be two distinct sizes in the modern era, particularly since the original unisex version was now rather small in an epoch that was increasingly moving toward the trendy 47 mm mark.

It didn't take Wunderman long to decide that the movement needed revamping. (Please see the next chapter for details on how this singular movement was updated for a twenty-first-century watch industry.) "He made it bigger and better," Orloff sums up. "And he launched his limited editions. We were crazy: we did it in platinum; we did it in rose gold; we did it in yellow gold; we did it as several series of twenty-five pieces. . . ." Orloff also recalls that Wunderman would insist on making prototypes of every single limited edition, whether there were orders for it or not. "He needed to see it, physically and tactilely, to be able to know that that was going to be something that he would want to release. So, unlike the Swiss, he kept making samples. And then, oh my God, he put diamonds on the Golden Bridge; this was unheard of."

"He touches stuff and it turns to gold."

The relaunch of the Golden Bridge came at a fortuitous moment in the brand's history: 2005 was simultaneously the half-century mark for Corum, and Wunderman wanted to celebrate the moment in his own style. "Severin was a friend of [celebrity sushi chef] Nobu; he used to babysit his kids," Orloff replied in answer to the question regarding the legendary party that Wunderman threw at Nobu's restaurant in Las Vegas in June of 2005 to celebrate the rerelease of his beloved Golden Bridge and Corum's first half-century. "He came out himself and cooked for Severin for that party." The party was gold, with gold food and women bodypainted in gold. The Stray Cats' Lee Rocker (Drucker) performed; this wasn't a stretch as Wunderman's daughter Debbie was married to the musician.

"He was truly visionary and it was a good thing and a bad thing. Sometimes he was way too ahead of the times."

"He was truly visionary," says Orloff, "and it was a good thing and a bad thing. Good because as a visionary he touches stuff and it turns to gold. And negative because sometimes he was way too ahead of the times. And he was this American guy, even though he was Belgian by birth—an American with an American mentality, in Switzerland trying to relaunch a brand and make it work on a global market. You know, Americans in Switzerland . . . he challenged everybody. When he had a vision, and he believed, he worked it, drew it, slept it, breathed it. I'd constantly get calls at 3:00 a.m."

FATHERS AND SONS

When Severin arrived at Corum in January of 2000, his son, Michael Wunderman, was in tow as the appointed vice president of marketing. Michael, the youngest of Wunderman's four children, schooled in Los Angeles and England, was at the time twenty-four. And, having worked with his father at Severin Montres for Gucci, he was the heir apparent. He moved to Switzerland to be closer to the factory and was named president in 2004.

Wunderman hired Antonio Calce after he'd left his position as general manager of Panerai in 2005 to fill the position of vice president of operations, and the three formed yet another triumvirate, jointly establishing strategies for global growth while rejuvenating traditional models such as the Golden Bridge and the Admiral's Cup. In 2007, Calce advanced to general manager when Michael returned to his native California to take over as the U.S. brand president.

Severin and Michael Wunderman

On June 25, 2008, Severin Wunderman passed away, at which point Michael advanced to creative director of the board. Serge Weinberg was appointed chairman of the board. A personal friend and financial adviser to Severin Wunderman, Weinberg was the former chairman of the management board of the Pinault-Printemps-Redoute (PPR) Group, today known as Kering. The Severin Wunderman Trust was created with Severin's shares bequeathed to it, to be directed by Richard Tomlin, while Serge Weinberg and Michael Wunderman were the other two members of the board.

In October 2009, Montres Corum Sàrl acquired Corum USA LLC, and Michael Wunderman left the watch industry. He continues to live in Los Angeles and now works as a producer in the film industry.

Opposite: Variations of the new Golden Bridge, mid-2000s

CORUM
CORUM

NEW BLOOD FOR A BRIGHT NEW FUTURE

Right from the beginning, the Severin Wunderman Trust was set up to continue investing in Corum only for a certain amount of time. Thus, it was clear that the brand would soon change hands once again, allowing for the advent of a new future to come over the horizon.

On April 24, 2013—the first day of Baselworld 2013, in a year when the fair had just undergone a complete, sensational architectural transformation—Corum released news that was almost bigger than the many multimillion-dollar booths now on display: it had been purchased by China Haidian, which was slowly but surely building a small group of watch brands. In addition to two mid-range Chinese watch brands, China Haidian had recently acquired Eterna.

In an interesting twist of fate, China Haidian Holdings Ltd, a company listed on the Hong Kong stock exchange, purchased 45 percent of the capital of Montres Corum Sàrl with China Haidian shares. This means that the Severin Wunderman Trust holds 10 percent of China Haidian, making it the second-largest single shareholder, right behind Kwok Lung Hon, who holds 67 percent of China Haidian. Therefore, in a certain way, Severin Wunderman remains tied to Corum, even in the afterlife.

China Haidian Holdings Ltd changed its name to Citychamp Watch & Jewellery Group Limited as of July 2014. The group's declared focus is on watches and jewelry. Brands it now owns include Corum, Eterna, and Rotary as well as Rotary's sister brands in the British-Swiss Dreyfuss Group Limited (which Citychamp also acquired) and Rossini and Ebohr, two of China's four major watch brands. Citychamp also distributes nonproprietary international brands in major cities across mainland China, making for a total of more than twenty-five brands. Additionally, the group owns about 280 retail shops and distribution outlets. "The new name marks a new milestone in our development, providing a clear new corporate image and reflecting our core business and future business direction," Hon said in a press release. "Looking ahead, we are aggressively expanding our watch and timepiece business, while leveraging our abundant experience in the design, production, and sales of watches and timepieces to expand into the jewelry business, inject new momentum into our development, and generate promising returns for our shareholders."

Today, Corum has returned to many of its original values and is refreshing original design ideas. Both Vincent Calabrese and Jean-René Bannwart remain in the Corum orbit as advisers.

Opposite: A white gold, diamond-set Golden Bridge from about 2005

A fully diamond-set Golden Bridge from the mid-2000s

Opposite: A fully diamond-set Golden Bridge after the case refresh

CORUM

V

BUILDING A BRIDGE
TO THE TWENTY-FIRST CENTURY

BUILDING A BRIDGE TO THE TWENTY-FIRST CENTURY

A biannual missive sent to Corum's retail, agent, and "Corum family" network in October 1980 proudly stated, "After long months of perfecting, the first little series of Golden Bridge movements with its 18-karat gold bridges were entirely fabricated within our workshops: Corum is for the first time a manufacture." In case the reader is unfamiliar with the sense of this word as it is used here, it comes from two Latin terms meaning "hand" and "make," and in watch jargon it signifies a company able to make its own movements. In the watch industry, which is at heart a cottage industry, it is normal for the individual components of a watch to be produced by expert suppliers, a manner of manufacturing that ensures the best quality of a product.

"Corum is for the first time a manufacture."

Despite this supplier tradition—which Corum generally subscribes to—manufacturing the Golden Bridge inside Corum's own walls was an important step for the brand on the road to becoming the creative watchmaker it is today. It allowed the company to take control of a process that it needed to control in order for the Golden Bridge to become a viable standing element of the brand's collection. Additionally, as Jean-René Bannwart remembers, "Little by little, people would say to us, 'Why don't you have your own movements?'" Thus, the patented Calibers 13 and 13 P were born. Caliber 13 P was a slightly modified version containing an extended main bridge for use in a pendant watch.

Even more important, the novel technical idea that was the base of Calabrese's award-winning invention, which comprised putting the time setting and winding stem right through the barrel containing the mainspring—done to ensure that there was no need for a crown to be in a visible spot—was a very delicate construct. This meant that winding had to be done very delicately without applying great force. All too often, timepieces were returned to the factory because owners had wound their watches much too vigorously. The mainspring was particularly delicate because the spring barrel had to be so small to retain the overall in-line look of the movement. The space available was too modest to contain a slip spring, which prevents overwinding.

As time progressed, it became clear that consumer demand for reliability and viability had changed, and Corum, now under the ownership of Severin Wunderman (see previous chapter), had to change along with it. Thus, Corum searched out a new mode of manufacturing the Golden Bridge that made sense for a new millennium.

Top: Assembling the new Corum Golden Bridge

Above: A peek inside the *haute horlogerie* workshop at Corum (photo Ian Skellern)

Corum Caliber CO 113: the revamped and refreshed Golden Bridge movement ready for the twenty-first century

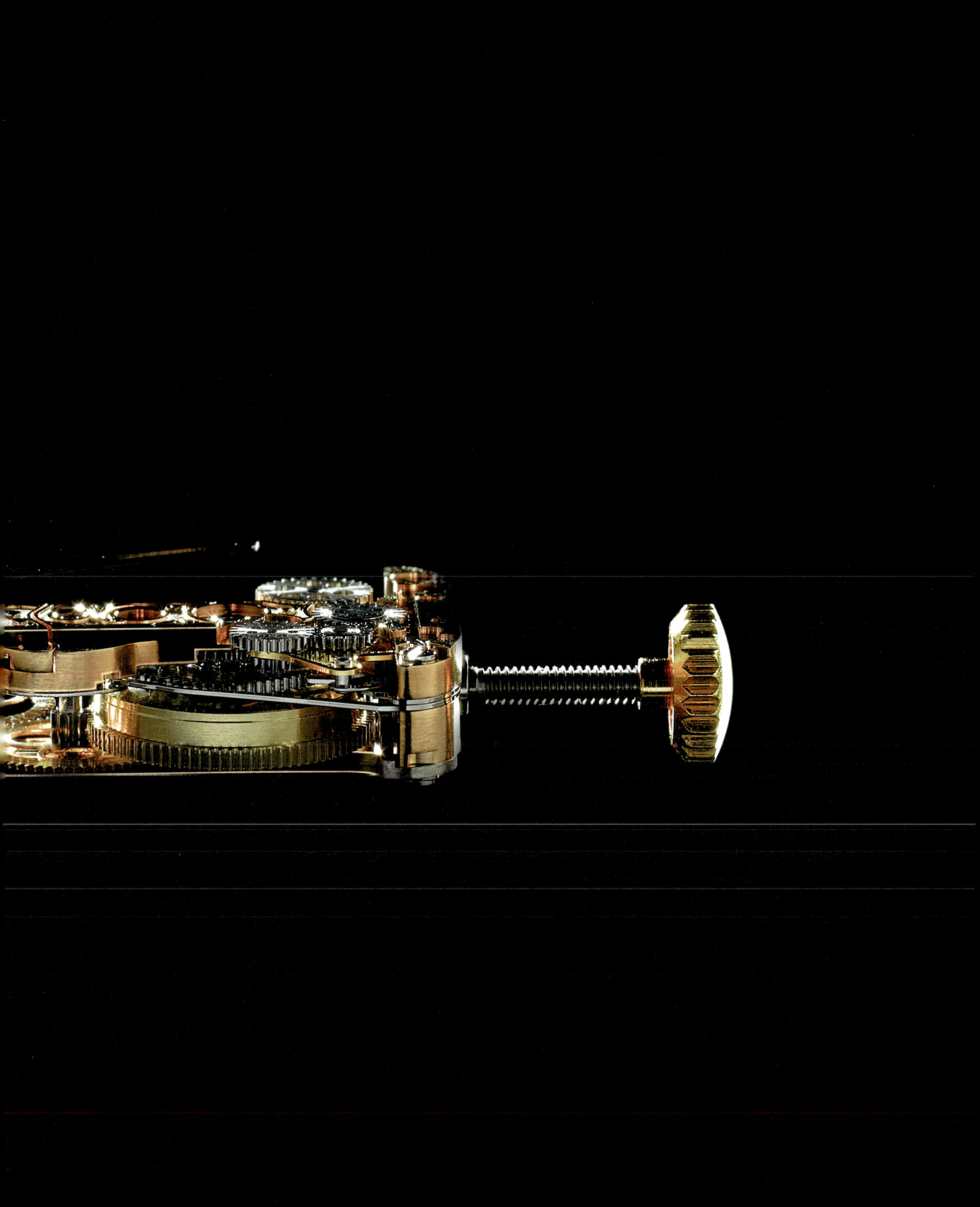

The Corum Golden Bridge Automatic in a titanium case with black DLC coating and a metallized sapphire crystal

Vaucher Manufacture Fleurier was established in January 2003 as the result of the official splitting of Parmigiani Mesure et Art du Temps SA into two entities: Parmigiani Fleurier, which concentrates on eponymous brand activities, and Vaucher Manufacture, which produces movements. A very short time later, Vaucher entered into a strategic alliance with Corum, which did not have a technical drawing team on staff, making the Bannwart-founded brand one of the very first of an exclusive list of illustrious clients that also includes Richard Mille (another artistic, forward-thinking brand that was founded in 2001). In 2006, Hermès acquired 25 percent of Vaucher.

One of the main changes carried out in designing new Caliber CO 113 was to move the winding stem from the back of the watch to the 6 o'clock position.

The development of the new Golden Bridge movement took place between 2003 and 2005 and was spearheaded by Sébastien Jeanneret on the Vaucher side. "We needed the full three years to develop, realize prototypes, validate, and homologate the final series for delivery to Corum," he remembers. "Corum's objective was to reach a high level of commercial success with the Golden Bridge. The original version had an excellent design and was successful in itself, but it had a few technical problems that made it impossible to add a tourbillon to it, for example." (See chapter 7 for more on the tourbillon version of the Golden Bridge.) One of these major technical deficits was that the winding stem was positioned underneath the movement, so that in order to repair or service it, the entire movement had to be dismantled. It also did not do very well with shocks.

Corum's goal was to make a modernized, "industrialized" movement—not industrial as in commonplace, but as in more industrial-strength and "normed." The idea was to continue producing small series, but series that could hold up to the needs of a contemporary, new-millennium wearer. The greatest strength of Calabrese's in-line movement design was effective energy transmission thanks to the placement of the gear train wheels all in one row, though this was indeed difficult for the watchmaker to assemble. According to Jeanneret, this was a most effective manner of efficient energy transmission.

One of the main changes carried out in designing new Caliber CO 113 was to move the winding stem from the back of the watch to the 6 o'clock position. This solved the problems of not being able to get to it for repairs (the winding stem is one of the weakest points of a wristwatch altogether and frequently needs servicing) and of the weakness derived from having to go through the spring barrel. Putting the crown between the lugs at 6 o'clock did not disturb the design or the watch's overall quest for transparency. In fact, this solution has become rather popular for later wristwatches with unique designs.

Adding hands to Caliber CO 113

Opposite: A mid-2000s Golden Bridge

CORUM

The placement of the crown so that its stem avoids running through the spring barrel is the principal difference between the original Golden Bridge movement and the freshly designed one. But another element, whose implementation was the result of trial and error, would become more important than immediately perceived: pillars placed along the sides of the movement. At Vaucher, the technicians did not anticipate having to use them, so the first prototypes were pillarless, so to speak. Adding pillars to the movement, however, proved important for shock resistance. During shock testing without them, the soft, 18-karat gold bridges separated. To ensure they remain in their designated places and to maintain the integrity of the sandwich construction, the technicians connected them with pillars.

Putting the crown between the lugs at 6 o'clock became rather popular for later wristwatches with unique designs.

A third important modification was adding the slip spring inside the spring barrel, which was not included in the first edition of the Golden Bridge movement, as the spring barrel was just too small to accommodate it. It was an elegant movement, and considering the level of energy needed to operate the time display, the volume of the barrel simply didn't leave any room for another element.

Compare this with a larger manually wound movement: when the mainspring inside the spring barrel is wound by turning the crown, one notes when the end is reached because a force seems to push back. That is the slip spring, which prevents overwinding. Inside a small, fragile movement like the original Golden Bridge there is no room for this clever little spring, and if force was used, it was likely to break the mainspring, the stem, or the crown. So the Vaucher technicians had to ensure that the new Golden Bridge movement could fit a slip spring to avoid this common problem.

Two examples of Caliber CO 113 inside first-edition cases featuring the stepped design: one in rose gold and one in white gold with diamonds

HIGHER FREQUENCY, MORE PRECISION

A watch's frequency is measured by the number of semi-oscillations a balance spring makes over the span of one hour (vph, or vibrations per hour), which can also be expressed in hertz. Hertz (abbreviated as Hz), named for German physicist Heinrich Hertz, is the unit of frequency defining this periodic phenomenon. It is calculated by dividing the number of oscillations by the period of time in which they are measured. A watch oscillating at a frequency of 18,000 vph (9,000 full oscillations) beats at 2.5 Hz (9,000 divided by 3,600 seconds equals 2.5). Until about the 1950s, wristwatches were mainly set to a frequency of 18,000 vph; since then, frequencies have steadily been rising, with 28,800 vph (4 Hz) having become the modern norm as a good precision standard.

The new assembly showed a 30 percent gain in regulation precision.

Another improvement to the Golden Bridge Caliber CO 113 is the balance's higher frequency of 28,800 vph (4 Hz) compared to the 21,600 vph (3 Hz) of the original movement, which was certainly—with exceptions, of course—the highest frequency in general use in the late 1970s. The new assembly, which also utilizes a variable inertia balance solution (as opposed to the traditional screw balance of the former), showed a 30 percent gain in regulation precision, according to Jeanneret. Also, because of the improved spring barrel, the new Golden Bridge boasts more than 40 hours of power reserve as compared to about 36 hours for the old one, despite the higher frequency.

It was important to have the right dependable, suitable elements in order to be able to pass classic watchmaking tests like those needed to attain the Chronofiable certificate. All of the system's elements, even the crown and stem, were tested for validity. Vaucher ensured that the new Golden Bridge movement was able to cruise through every test possible. Attaining a COSC certificate, though now technically an option, was not achievable because the Golden Bridge does not have a second hand. Such an addition would naturally destroy the careful design of the piece.

Utilizing modern materials and methods also raised the overall reliability of the movement, as did the redesign, which slightly altered the original straight-line shape to form a T on both ends of the main bridge (due to the redesigned case, this is not obvious from the outside). This symmetry not only allowed for better dispersion of the energy and tension, but also made for more stability in case of shock and more rigidity to prevent that shock.

Today, the Vaucher-manufactured kits come to the Corum factory in La Chaux-de-Fonds for assembly in the "atelier": the workshop dedicated to the assembly of the Bridge timepieces and other specialties like the Coin Watch. The new Golden Bridge movement is certified by the independent Swiss organization Chronofiable as being mechanically sound and reliable.

Above: Careful assembly in Corum's dedicated "atelier," where the *haute horlogerie* timepieces are put together

Right: Special benches in the "atelier" ensure a dust-free environment for assembly (photo Ian Skellern).

CORUM
18K

Corum Caliber CO 113

A MORE SOLID BRIDGE CASE

As we learned in chapter 3, the transparent case is just as important to the success of the Golden Bridge as the clever movement. Naturally, then, the new case was just as important to the revamped model as the redesigned movement. François Senderos is a longtime employee forming part of the Corum team that developed the new Golden Bridge designed by Stefano Minello, who had by then replaced Schwab, and was tasked with spearheading the new casing.

The new design made the casing strong, able to withstand much greater shock, and accentuated the straight line of the movement by echoing it.

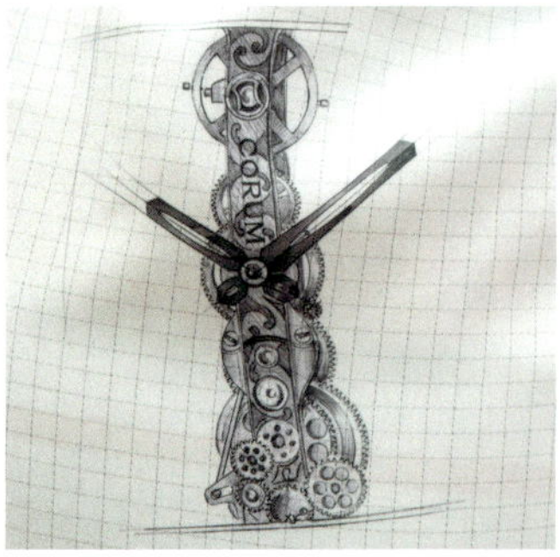

Left: The "atelier" is home to Corum's most talented watch-makers.

Above: The Golden Bridge movement is proudly displayed on the technicians' white coats (photo Ian Skellern).

"The new case was designed completely differently than the cases housing the first movement," Senderos remembers. The new case was now offered in sizes targeted for the male and female wrist respectively and contained four crystals: one on the back, one on the front, and one on each lateral side. In between all the crystals runs a shaft of gold or platinum. This design made the casing strong, able to withstand much greater shocks, and accentuated the straight line of the movement by echoing it. Producing and assembling the entire construct was also much easier. The case underwent a slight refresh a couple of years later that reduced the stepping at the top and the bottom, making it a bit lighter-looking and more minimalist. For Baselworld 2013 it was also released with a metalized back crystal for less transparency in accordance with men's watch trends, but the principle has remained the same.

FIRST LIMITED EDITIONS

The new Golden Bridge relaunched at a celebration held in June 2005 in Las Vegas (see chapter 4 for more details). The first wristwatch to emerge was a limited edition of 200 total pieces (fifty in white gold, fifty in yellow gold, fifty in red gold, and fifty in platinum) celebrating fifty years of Corum and, naturally, the rebirth of one of its most well-known and luxurious timepieces.

This extravagant edition was not fully transparent as it featured a dial with numerals with an ivory-colored lacquered surface. The numeral at 5 o'clock was decked out in diamonds to represent "50." Obviously, it was housed in the first of the new Golden Bridge cases, which was quite solid-looking and stepped.

A little later that year, Corum donated a one-of-a-kind Golden Bridge to the second edition of the now-biannual Only Watch auction, which took place on September 22, 2005. Proceeds of the unique auction held in Monaco during the Yacht Show benefit the Association Monégasque Contre les Myopathies (Monegasque Association against Muscular Dystrophy). It was conceived and organized by Luc Pettavino, whose son is afflicted with the rare Duchenne strain of muscular dystrophy. It has become a highly anticipated event in the watch world, and Corum, an early patron, has continued to donate a watch to every one of the charitable auctions.

A new Golden Bridge for 50 years of Corum: limited to just 50 pieces each in white gold, yellow gold, red gold and platinum

The watch submitted in 2005 was housed in an 18-karat white gold case measuring 40 x 32 mm. The visible movement was set against the backdrop of a hand-painted miniature featuring two posturing stallions originally painted by Swiss artist Hans Erni. The horses positioned on either side of the highly visible Caliber CO 113 remain completely legible thanks to the use of skeletonized white gold hands. Reflecting Severin Wunderman's love of art (see the previous chapter), the watch was donated along with a lithograph of the original artwork depicted on the dial. This unique piece's auction estimate was listed as €20,000 in the accompanying Antiquorum catalogue, but it ended up selling for more than double that at €50,000.

A new Golden Bridge for 50 years of Corum: note the diamond set at 5 o'clock on the dial making it look like "50"

Opposite: The one-of-a-kind Golden Bridge for Only Watch 2005

CORUM

WHAT MADE THIS MOVEMENT SO SPECIAL?

Before the Golden Bridge, the general ideal for timepieces was to put precision and reliability on display, not mechanical artistry. The approach of the Golden Bridge certainly changed the way this was generally viewed as this timepiece's main purpose was to unite design and aesthetics with mechanical functionality. Previously, design was relegated to the elements surrounding the movement, and the movement was something functional that was integrated in one piece into the design. The aim of the Golden Bridge was to be able to see the movement, like a little jewel inside a transparent cushion. The Golden Bridge is also evidence that mechanics and design can be completely aligned in every sense of the term. The functional aspect of this movement is to put the design first, to be something original but technical at the same time. And this perfectly expressed the original spirit of René Bannwart.

Caliber CO 113: the redesigned Golden Bridge movement

TECHNICAL SPECIFICATIONS OF THE FIRST NEW GOLDEN BRIDGE

The first serial Golden Bridge of the new era as introduced at Baselworld 2006

MOVEMENT	Baguette shape, 33 x 3 x 4.9 mm Hand wound via crown at 6 o'clock Corum designation CO 113, Vaucher designation V700 VMF Hand-engraved main plate and bridges in 18-karat white or rose gold Variable inertia balance adjustable by 4 inertia blocks in 18-karat gold 40-hour power reserve 140 individual components including 19 jewels
FUNCTIONS	Hours, minutes
CASE	Tonneau-shaped 18-karat white gold, set with 66 diamonds (1.26 ct), rose gold or platinum Sapphire crystals on the top and sides Transparent screwed-in sapphire crystal case back
CRYSTAL	Four specially shaped and cut sapphire crystals
WATER RESISTANCE	Splash-resistant (later became water-resistant to 3 atm in newer case)

CORUM

VI

GOLDEN BRIDGE LIFESTYLE

EN VOGUE

While it's not a fashion brand, Corum has always been fashionable. This is not to say that the the brand's watches have matched fashion trends, for example the fleeting fashion of the catwalk, but rather the watches—well ahead of their time—have been both enjoyable to wear and conversation pieces like any good piece of jewelry or notable item of clothing.

The Golden Bridge, in particular, has always attracted very fashionable wearers—despite its timelessness. Now in its fourth decade of existence, the Golden Bridge still looks just as fresh as the day it was born.

This might be the reason that fashion photographers of world renown have set this wristwatch in scene over the years. It is more than an object or a prop: the Golden Bridge has personality, and the looks and brains to match. The models wearing it may be the props for it!

While it's not a fashion brand, Corum has always been fashionable.

Left: This Golden Bridge is an original from 1980.

Opposite: She is wearing the Golden Bridge Dragon from 2014, while he wears a modern Golden Bridge in red gold.

Golden Bridge Automatic

He sports the modern Golden Bridge in white gold.

The modern Golden Bridge in a ceramic case with red gold crown

CORUM

The romance of the Golden Bridge's mechanics has an irreversible effect on the senses.

Above: A diamond-set Golden Bridge with blue patent leather strap.

Opposite: A modern Golden Bridge set with baguette-cut diamonds and rose-colored sapphires.

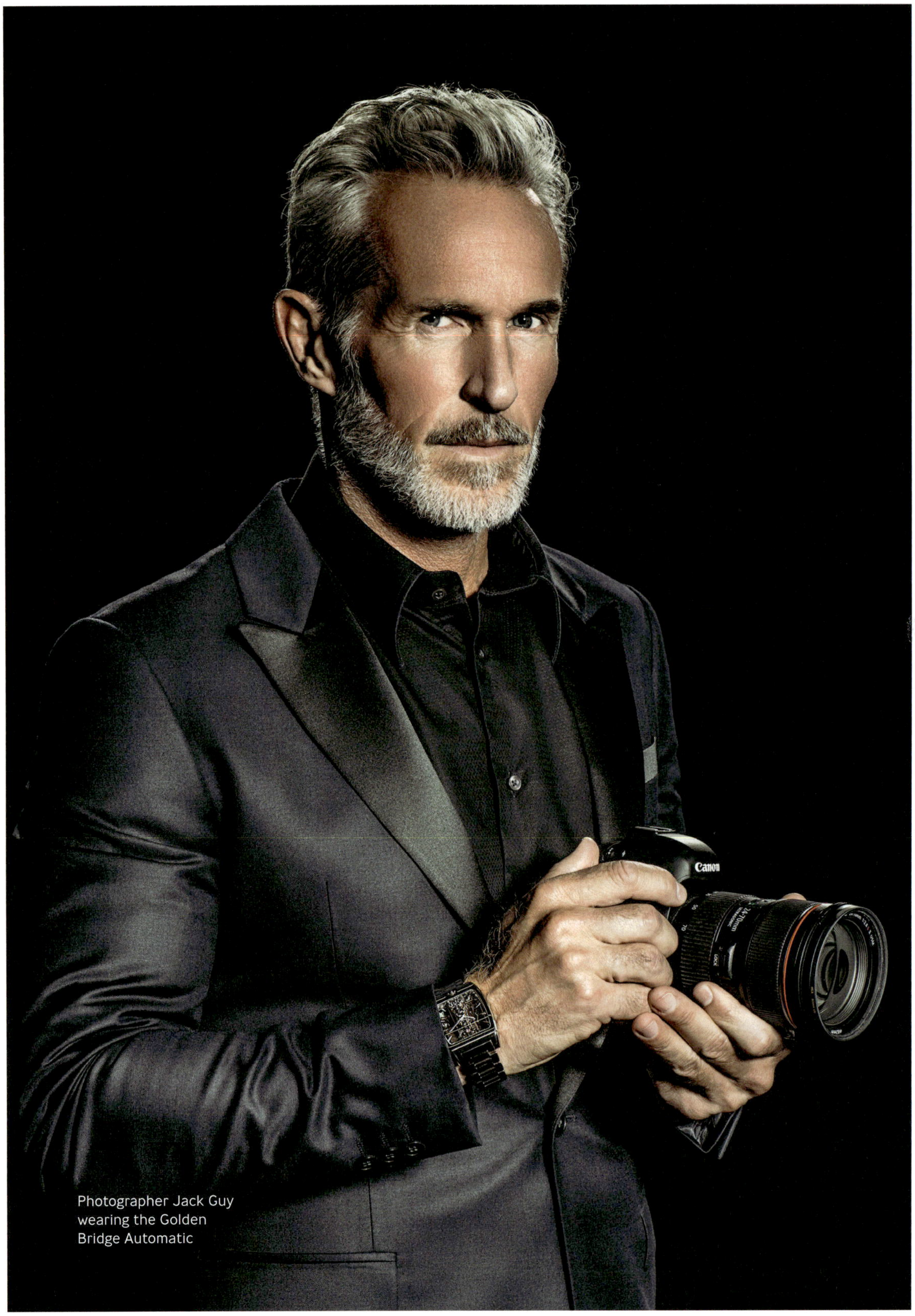

Photographer Jack Guy
wearing the Golden
Bridge Automatic

The modern Golden Bridge set with baguette-cut rose-colored sapphires on a matching patent leather strap.

Modern Golden Bridge
in red gold sporting 180
diamonds on the bezel

Modern Miss Golden Bridge in red gold boasting 90 diamonds on the bezel

The Miss Golden Bridge
in rose gold

The original Golden Bridge from 1980

A Golden Bridge variation
from the 1980s

It's only rock and roll: the original Golden Bridge from 1980

THE CRYSTAL BACCARAT GOLDEN BRIDGE CLOCK

René Bannwart's first big sales successes were in France through renowned retailers such as Cartier and Van Cleef & Arpels. Surely this is what prompted the relationship between Corum and Baccarat, a traditional crystal maker located in the Lorraine region of France. Baccarat, founded in 1764, is one of France's most famous brands. Together, the two artistic companies created the Baccarat Golden Bridge crystal clock, which captured a larger version of the unique movement inside a sheath of French crystal and planted it squarely on a plinth of 18-karat gold. Each element of these clocks was completely made by hand. The clocks were conceived as eye-catchers for store windows, and there are no more than 300 pieces in total in existence.

Each element of these clocks was completely made by hand.

As each element of the Golden Bridge Baccarat Clock was made by hand, each clock was basically unique.

HAND IN THE POCKET

The Golden Bridge was, and remains, as much a statement of personal style as a proclamation that one is "in-the-know" when it comes to unique, quality horology. Indeed, it is the superpower of this particular timepiece to combine the best of all possible worlds when it comes to wearable items.

Corum was often ahead of its time when it came to artistic ideas. Thus, Corum was also one of the first watch brands to advertise in magazines like *Vogue*, but also technically minded periodicals like *Scientific American*. However, it was the Corum advertising campaign called *"main dans la poche"* ("hand in the pocket") that was most characteristic of the brand. Today, we might call it a wristshot, but in a time predating mobile phones and selfies, this photographic view of a wristwatch—on an actual wrist—was not all that common. It remains a true eye-catcher.

A notable idea used in Corum's advertising throughout the decades: "hand in the pocket" is probably the world's first example of a wristshot.

VII

THE GOLDEN BRIDGE FAMILY

THE GOLDEN BRIDGE FAMILY

Corum is a company that has built its reputation on making art for the sake of art. This means that almost every single timepiece starts with an emotion. This aesthetic of risk—after all, art is subject to the whims of personal taste—therefore connects on a deep level to the individuals who buy the brand's timepieces. The romance of the Golden Bridge's mechanics, even if they are not fully grasped in a technical sense, has an irreversible effect on the senses. This is not a machine; it possesses something that evokes very human emotions.

Making art for the sake of art.

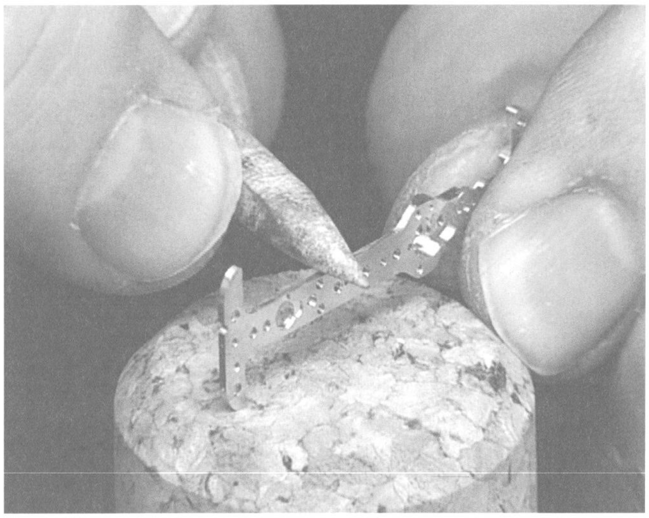

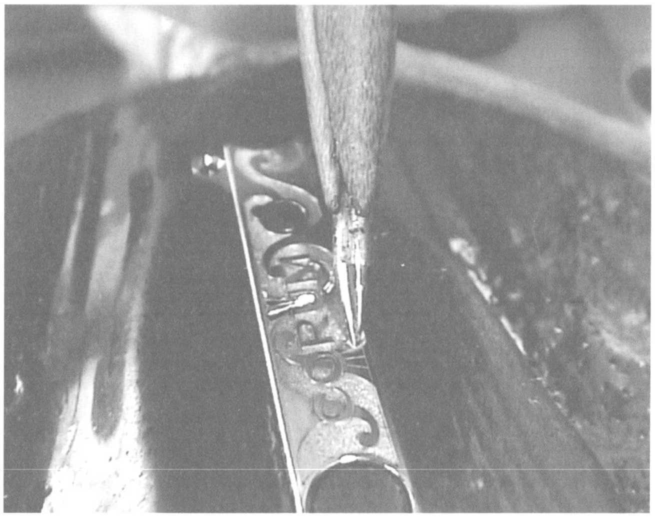

The management team led by Antonio Calce decided it would make sense to have a full line of Golden Bridge family members, so one of the things that has characterized the company's recent history was the challenge to add complications to the Golden Bridge as well as creating new variations on the "bridge" theme.

Left: Beveling the sides of a Golden Bridge main plate

Right: Engraving a main plate of the Golden Bridge's Caliber CO 113

Opposite: Adding the oscillating weight to the Golden Bridge Automatic

A NEW SUBFAMILY: THE TI-BRIDGE

The first of these to see the light of day was the Ti-Bridge, which launched at Baselworld in 2009. And while it may look like the Ti-Bridge was simply the Golden Bridge turned sideways, things weren't quite that simple. First off, Caliber CO 007 was created in titanium in collaboration with movement maker La Joux-Perret and not Vaucher. It was not executed in gold (hence the name), which meant that a whole new development was necessary as titanium is very difficult to machine. Second, it had a much larger spring barrel designed to achieve the 72-hour power reserve. Finally, it was secured to the case by crossbars, which made for an interesting, modern look that also provided a greater cushion against shock. The automatic version, Caliber CO 207, which utilized two oscillating weights and was called the Automatic Dual Winder, followed in 2013.

Despite its generous dimensions, the titanium watch wore very ergonomically, almost as lightly as air.

The horizontal baguette tourbillon version launched in 2010 was also created in conjunction with La Joux-Perret. It, too, mimicked the unique look of the legendary Golden Bridge movement; Corum Caliber CO 022 appears to be almost like the original mechanism conceived by Vincent Calabrese turned on its side. Its manually wound movement was crafted using a few new materials such as ARCAP in addition to titanium and utilized double spring barrels to achieve the 72 hours of power reserve. However, its balance seemed slightly off due to the tourbillon set to one side at its left end covered by a steel Corum key logo acting as a balance cock, which was secured by two large screws. Despite its generous dimensions, the titanium watch measuring 42.5 x 41.5 mm wore very ergonomically, almost as lightly as air.

The Ti-Bridge Lady was a very decorative and elegant new addition to the family, paring the Ti-Bridge look down to an elegant rectangle and incorporating a few trendy elements like the use of the color white, novel ceramics, and, audaciously, incorporating diamonds.

It was obvious that Corum was not looking to emphasize the original transparent visuals of the Golden Bridge with the advent of the Ti-Bridge, but rather to extend the Bridge family.

FOR THE LADIES

The year 2009 also marked the experiment known as the Golden Bridge Lady. Unlike the Miss Golden Bridge, which launched with the refreshed movement and case, this particular timepiece actually included the Golden Bridge movement turned on its side, allowing the case to be radically enlarged to 41 x 34 mm and thus making it feel more contemporary while still fitting perfectly due to the fact that it sits lengthwise on the wrist. It came in red or white gold and was set with a fireworks display of diamonds weighing in at about 2.5 ct.

The Golden Bridge Lady (left) and the Miss Golden Bridge; the latter, outfitted with 233 diamonds and rose-colored sapphires, is still in the current collection.

The Golden
Bridge Tourbillon

A WHIRLWIND AND A FIRST

The year 2010 marked a dual anniversary for Corum: thirty years of the Golden Bridge and fifty-five years of Corum. For this reason there were a large number of new releases at Baselworld, and among them was a limited edition wristwatch that marked a first in two ways.

2010 marked thirty years of the Golden Bridge and fifty-five years of Corum.

The Golden Bridge Tourbillon included not only a tourbillon that contained what is quite possibly, at 8.5 mm in diameter, the world's second smallest tourbillon carriage, but also constituted a first by having a silicon escapement nestled within. Caliber CO 213 was comprised of 182 components all strung along the miniscule baguette-shaped bridge measuring 33 x 3 mm. The silicon was used for the three escapement components, which not only allowed the reduced sizing, but also their ability to function perfectly without lubrication. The balance, which was specially made by Vaucher to accommodate the changes, beat at the unusual frequency of 19,200 vph (right between 3 Hz and 4 Hz to get the best of both worlds in terms of precision while going easy on the power reserve), and despite the smaller spring barrel, it still boasted 40 hours of power reserve.

It goes without saying that combining this ultramodern escapement with the more traditional 18-karat gold bridges hand-engraved with patterns of tree ferns found in the Jura region makes for a wonderful juxtaposition. The cage of this tourbillon, crafted in lightweight titanium, was covered by a steel Corum key acting as a balance cock.

The first version of the tourbillon was launched in an edition of thirty-three pieces (sixteen in red gold, eleven in white gold, and six in platinum). Because the technicians at Corum and Vaucher wanted to keep the mechanism as original as possible, on this very first trial, as on the Ti-Bridge Tourbillon, the key-shaped steel cock unfortunately obscures some of the view of the rather small revolving whirlwind due to the gold bridge. The latter had already been made a bit thinner, but part of the tourbillon was still hidden. To solve this—most aficionados would rather have an unobstructed view of the tourbillon—the technicians later hit upon the idea of making a sapphire crystal bridge that one can see through. But they couldn't allow the sapphire crystal to be attached on both ends as that would have too many points of tension and could break (sapphire crystal is rather brittle). So the final solution ended up with the bridge fixed on one side with a screw, while on the other side it has a sapphire crystal piece shaped appropriately, allowing it to withstand the tension placed on it by the screw. It goes without saying that this is an exceptionally creative solution that alludes to the history of the line while definitively bringing it forward into the present.

Every single component had to be hand-adjusted to the tolerances of the next one, which helped in terms of measurements and power reserve.

Adding a tourbillon also meant that there was extremely limited space to work with. So, the tourbillon had to be very small. In order to accomplish that, every single component had to be hand-adjusted to the tolerances of the next one, which helped in terms of measurements and also in terms of power reserve. Because there is less play between components, there is less energy lost due to undue friction. Nonetheless, the technicians still could not fit in the same number of components plus the tourbillon for the 33 mm mechanism. To solve this, they placed the third wheel off to the side. However, when they were done with development, they realized that moving the third wheel to the side gave the Golden Bridge Tourbillon a lopsided look. The solution? They added a second third wheel just for aesthetics: it turns along with the rest of the gear train but with no function.

The movement of the Automatic Golden Bridge: Caliber CO 313

AUTOMATIC WINDING FOR THE GOLDEN BRIDGE

In 2011 the Golden Bridge received an automatic sibling: Caliber CO 313. While for any other timepiece, this would not be grounds to celebrate, but instead to ask why it took so long to bring it to market, it goes without saying that adding an automatic winding to the Golden Bridge's svelte baguette-shaped movement without radically altering its elegant and transparent look is nothing short of a mechanical wonder.

Requiring four years of development and testing—the difficulty of its development is easily illustrated by the fact that the tourbillon was released one year before the automatic winding version—the challenge put to the Corum and Vaucher teams consisted in maintaining the linear essence of the movement despite adding the comfort of a self-winding mechanism. To keep the movement's design pure, coming up with a unique solution had to be the order of the day. And thus the linear winding system was invented—something that had not yet existed in the world of mechanical watches in quite this form.

The oscillating weight is crafted in platinum, and thanks to the inherent transparency of the Golden Bridge timepiece, it is visible from both sides of the case. Naturally, the technicians learned a great deal from developing Caliber CO 113, so the movement, which beats at 4 Hz, also houses the slip spring within the spring barrel to ensure that overwinding does not occur, even if it's not a human hand doing the winding all the time. Containing just 194 individual components, including a variable inertia balance for long-term precision, it remains a minimalist example of creative horology.

The linear winding system had not yet existed in the world of mechanical watches in quite this form.

The oscillating weight, at 4 grams, is crafted from a dense, heavy metal that is also one of the most precious on earth. It moves freely along rails positioned on the back of the movement in a radius of just one centimeter, after which its motion is dampened by a system of gaskets including silicon joints that push it back in the other direction. The addition of the silicon joints ensured that the platinum weight did not crash too loudly or heavily on the gaskets—not because it would hurt them, but because the wearer surely would not like feeling the rotor slamming against the other components all the time. This allows the wearer to feel that the weight is moving, but not in a distracting sort of way.

Friction was naturally a big issue for this new invention. The technicians tried out ten different material combinations before settling on cupro-beryllium for the carriage of the oscillating weight and steel with a nickel-polytetrafluoroethylene (PTFE) coating comparable to Teflon for the rails to prevent undue friction. The energy created from the motion of the linear rotor, which engages with a reverser system outfitted with ceramic ball bearings requiring no lubrication to ensure bidirectional winding, is transmitted via a rack.

The technicians tried out ten different material combinations.

One of the other main challenges was the winding capacity, which in the end boasted a more than 2:1 ratio: wearing the watch for one hour allowed it to wind for two hours. To achieve this, the movement designer had to maximize the length of the weight until the proper ratio was achieved.

The Golden Bridge thus retains all of its elegant appeal while receiving the comfort of automatic winding. It is available in limited quantities only.

DRAGON, PHOENIX, AND OTHER ARTISTIC PEARLS

As a company basing its collections on design elements, Corum has always been a risk-taker. This is illustrated by the number of one-off, special, artistic watches that has come about over the course of the last sixty years. These have defined the spirit of the company, whether it was René Bannwart or Severin Wunderman at the design helm.

The more recent rereleases of some of these special editions and unique pieces show that these artistic watches have not been forgotten. One very fine example is the Golden Bridge Dragon that came out at Baselworld 2014—followed up by the Golden Bridge Dragon and Phoenix at Baselworld 2015—which combines the unique baguette-style movement with a beautifully engraved, visually striking dragon winding its way around the movement balancing a pearl of wisdom on its preciously detailed tail.

Top: The Golden Bridge housed in a DLC-coated titanium case

Above: The Golden Bridge Dragon sees a mythical creature fiercely guarding its mechanical treasure.

TECH SPECS GOLDEN BRIDGE TOURBILLON OF 2011 (SECOND VARIATION)

MOVEMENT	CO 213 Mechanical baguette movement, hand-wound by crown at 6 o'clock Tourbillon with silicon escapement, sapphire crystal bridge Slip spring Variable inertia balance wheel Frequency 19,200 vibrations/hour (2.66 Hz) 22 jewels 18-karat red gold plates and bridge Power reserve of 40 hours
FUNCTIONS	Hours, minutes
CASE	34 x 51 mm Tonneau-shaped case made of 18-karat red gold Polished and satin-finished surfaces Case back engraved with Corum logo
DIAL	Dial in 18-karat gold, guilloché by hand Applied hour indexes Red gold skeletonized baton-shaped hour and minute hands
CRYSTAL	Sapphire crystal
WATER RESISTANCE	30 meters (3 atm)

TECH SPECS GOLDEN BRIDGE AUTOMATIC OF 2011

MOVEMENT	CO 313 Automatic baguette-shaped movement with linear winding Oscillating weight made of platinum Slip spring Variable inertia balance adjustable by screws Frequency 28,800 vibrations/hour (4 Hz) 26 jewels Plate and bridges made of 18-karat gold, guilloché decoration Power reserve of 40 hours
FUNCTIONS	Hours, minutes
CASE	37.2 x 51.8 mm Tonneau-shaped case made of 18-karat red gold or 18-karat white gold Sapphire crystal case back cover
DIAL	Black or gray Applied hour indexes Skeletonized baton-shaped hour and minute hands
CRYSTAL	Sapphire crystal
WATER RESISTANCE	30 meters (3 atm)

VIII

APPENDICES

"KEY" DATES (COMPANY CHRONOLOGY)

1955 Corum founded with three partners: René Bannwart, Gaston Ries, Juliette Ries

1958 Gaston Ries dies; Simone Ries replaces him

1966 Jean-René Bannwart joins Corum

1977 Vincent Calabrese's Spatiale wins a gold medal at the Geneva Invention Salon

1980 Golden Bridge introduced at the Musée International d'Horlogerie (MIH)

1987 René Bannwart retires; Jean-René Bannwart takes over

1995 Factory addition inaugurated

2000 Severin Wunderman buys 90% of Corum (Jean-René Bannwart retains the other 10%)

2001 Severin Wunderman acquires the remaining 10%

2004 Michael Wunderman becomes president of Corum

2005 The Golden Bridge celebrates its 25th anniversary with the reengineered Caliber CO 113

2007 Antonio Calce becomes general manager; Michael Wunderman becomes president of the U.S. subsidiary

2008 Severin Wunderman dies

2008 Severin Wunderman's shares are bequeathed to the Severin Wunderman Trust chaired by Richard Tomlin; Michael Wunderman advances to creative director of the board

2008 Serge Weinberg is appointed chairman of the board of Corum

2009 Caliber CO 007 (a horizontal linear movement with titanium elements for the Ti-Bridge) is launched at Baselworld

2010 The Golden Bridge Tourbillon launches at Baselworld; this is the first use of silicon in a movement at Corum

2010 René Bannwart dies

2011 Acquisition of Les Artisans Horlogers along with 20 employees

2011 The automatic Golden Bridge, Caliber CO 313, is launched at Baselworld

2013 China Haidian, with principal shareholder (67%) Kwok Lung Hon, takes over Corum thereby making it the second largest shareholder of China Haidian

2014 Corum sets up an executive committee to manage the company; China Haidian changes its name to Citychamp Watch & Jewellery

SOURCES AND BIBLIOGRAPHY

BOOKS

Corum: The Story of Creation by Gisbert L. Brunner, 1993
Custom published by Corum

The Hands of Time by Ian Skellern and Peter Speake-Marin, 2010
Custom published by the AHCI
ISBN 978-2-8399-0628-9

Masters of Contemporary Watchmaking by Michael Clerizo, 2009
Published by Thames & Hudson Ltd.
ISBN 978-0-500-51485-6

Omega Saga by Marco Richon, 1998
Published by Fondation Adrien Brandt for Omega
ISBN 2-88380-010-3

Omega Voyage à Travers le Temps by Marco Richon, 2007
Custom published by Omega
ISBN 978-2-9700562-0-1

Swiss Timepiece Makers 1775–1975 by Kathleen H. Pritchard, 1997
Phoenix Publishing for the National Association of Watch and Clock Collectors, Inc.
ISBN 0-914659-79-0

Die Zeit der Uhren by Lucien Trueb, 1999
Published by Ebner Verlag Ulm
ISBN 3-87188-009-4

MAGAZINES

ArmbandUhren 1/2005, pp. 64–77, "50 Jahre Corum: Design-Objekte mit Zeitanzeige" by Gerhard Claußen

Europa Star No. 3, 1963, pp. 60–66, "René Bannwart"

ONLINE

ACN Newswire, July 7, 2014
http://en.acnnewswire.com/press-release/english/17321/china-haidian-officially-renamed

Artinfo.com, "Los Angeles County Museum on Fire" by William Poundstone, June 7, 2011
http://blogs.artinfo.com/lacmonfire/tag/severin-wunderman/

Europa Star, October 5, 2005
http://www.europastar.com/news/1001260230-corum-s-golden-bridge-at-only-watch-2005.html

Europa Star, by Jean-Luc Adam, May 3, 2013
http://www.europastar.com/news/1004086191-the-background-to-china-haidian-s-acquisition-of.html

Justia Patents
http://patents.justia.com/assignee/corum-ries-bannwart-co
http://patents.justia.com/inventor/rene-bannwart
http://patents.justia.com/inventor/severin-wunderman

Legacy.com, obituaries, from The Los Angeles Times, June 29, 2008
http://www.legacy.com/obituaries/latimes/obituary.aspx?pid=112362824

ThePuristS, "ThePuristS Interview Severin Wunderman Chairman of Montres Corum SA," by Jaw, August 2003
http://www.tp178.com/tjn/corum/sw/sw%5B2a%5D.htm

The Star Online, by Kee Hua Chee, June 18, 2006
http://www.thestar.com.my/Story/?file=%2F2006%2F6%2F18%2Flifefocus%2F14527840&sec=lifefocu

WITH THANKS

No book is ever the creation of a single person; there is always a large outpouring of generosity and aid from others that help make any book a successful tangible object. In this case, it was a superb, almost impromptu brainstorming session and a number of congenial follow-up discussions that planted the seed of the idea that was to bloom faster than any of us would ever have thought.

The author would like to thank the following people for their precious help in making this book the informative and interesting work it has become.

Boon Soon Chong, **Marie-Alexandrine Leibowitch**, **Antoine Hastoy**, **Jacques-Alain Vuille**, **Jean-René Bannwart**, **François Senderos**, **Sébastien Jeanneret**, and, of course, **Claude Roulet** for their enthusiastic, insider knowledge of Corum, support of the idea, and the energy to turn it into reality. Without you, this book would never have been possible in any way, shape, or form.

Vincent Calabrese for believing in artistry and artisanal horology, the ideal of the AHCI, and allowing me to pick his brain as often as I like.

Fabrice Mugnier and **Suzanne Wettstein** of Watchprint for staying loyal to book publishing in a highly digitized age and for supporting precious projects like this one in the specialized field of horology.

Sherry Williams Pantieri for endless energy, artistic vision, and the will to always go that extra mile. **Jack Guy** for extraordinary photographs that pull us out of our conservative way of thinking and open us up to just that little bit more of something we never knew was there.

Ian Skellern, my dear business partner at Quill & Pad, for his continuing moral support and the codiscovery that sometimes that one more thing won't be the straw to break the camel's back. And for his extraordinary photographs and textual advice.

Ashley Benning, as always, for her detailed eye, impeccable mastery of the English language, and extreme professionalism.

Knych Keller, **Stacie Orloff**, and **Lisa Delane** for sharing their memories of **Severin Wunderman** and the U.S. market.

Robert-Jan Broer for his expertise in vintage Omega and wonderful photographs.

Gary Girdvainis for inadvertently planting the initial seed and once again proving that the glass is usually half full even when you might think it is half empty. But also for helping me kick-start my path oh-so-many years ago.

Jan Abellan for his enthusiastic and creative graphic skills that really iced the cake.

Bénédicte Vandelle and **Sabrina Dörr** for their extremely competent, patient, and fast multilingual transcription skills

And special thanks to my loving family, **Armin**, **Alexander**, and **Sabrina Dörr**, for understanding that for me this job is more than just a way to earn a living.

ABOUT THE AUTHOR

There are only a handful of people who do what **Elizabeth Doerr** does for a living, and only a miniscule fraction of these people are women. Doerr writes about exceedingly complicated, exceptionally expensive watches that are designed to appeal primarily to men.

Having dedicated most of her professional life to mechanical watches since 1991—the year she first embarked on the journey to the trade fair that is now known as Baselworld—Doerr remains a contributor to numerous print and digital publications all over the world even after having cofounded her own digital publication in 2013: Quill & Pad (QuillAndPad.com).

A native of Petoskey, Michigan, Doerr has lived in Germany since 1988, when she moved there to pursue graduate studies in German language and literature after graduating from Kalamazoo College with a bachelor's degree in German and European studies. She now resides on the northern edge of the Black Forest with her husband, two children, and Persian cat.

Known for her deep understanding of the technical aspects of watchmaking and her personal way of approaching watchmakers and corporate managers alike, her thousands of stories written over the years reflect a holistic, candid view of high watchmaking as an art form depicting horology as a cultural treasure as well as a fond obsession for so many. Her writing often reflects her own enthusiasm about her subjects.

Doerr regularly sits on industry juries and committees. These notably include the Steering Committee of the Fondation de la Haute Horlogerie (FHH), an organization called to life to preserve and support the message of high watchmaking, and watchmaking's "Oscars," the Grand Prix d'Horlogerie de Genève (GPHG). The journalists called to partake as jurors in the involved competition for likely the most prestigious event in the world of watchmaking are generally recognized as being highly influential in their chosen fields.

After coauthoring *Elite Timepieces* in 2008, Doerr wrote *12 Faces of Time*, an artistic look at independent watchmaking, which was published by teNeues in February 2010.